Terahertz Physics

Terahertz physics covers one of the least explored but richest regions of the electromagnetic spectrum. Designed for independent learning, this is the first textbook to open up this exciting new area to students of science and engineering.

Written in a clear and consistent style, the book focusses on an understanding of fundamental physical principles at terahertz frequencies and their applications. Part I outlines the foundations of terahertz science, starting with the mathematical representation of oscillations before exploring terahertz-frequency light, terahertz phenomena in matter and the terahertz interactions between light and matter. Part II covers components of terahertz technology, from sources of terahertz-frequency radiation, through the manipulation of the radiation, to its detection. Part III deals with applications, including time-domain spectroscopy.

Highlighting modern developments and concepts, the book is ideal for self-study. It features precise definitions, clear explanations, instructive illustrations, fully worked examples, numerous exercises and a comprehensive glossary.

R. A. Lewis is a Professor in the School of Physics and the Institute for Superconducting and Electronic Materials at the University of Wollongong, Australia. Lewis is an active researcher in the area of terahertz science and technology.

Terahertz Physics

R. A. LEWIS
University of Wollongong, Australia

CAMBRIDGE
UNIVERSITY PRESS

University Printing House, Cambridge CB2 8BS, United Kingdom

Cambridge University Press is part of the University of Cambridge.

It furthers the University's mission by disseminating knowledge in the pursuit of
education, learning and research at the highest international levels of excellence.

www.cambridge.org
Information on this title: www.cambridge.org/9781107018570

© R. A. Lewis 2012

This publication is in copyright. Subject to statutory exception
and to the provisions of relevant collective licensing agreements,
no reproduction of any part may take place without the written
permission of Cambridge University Press.

First published 2012

A catalogue record for this publication is available from the British Library

Library of Congress Cataloguing in Publication data
Lewis, R. A. (Roger Adrian), 1957–
Terahertz physics / R.A. Lewis.
   p. cm.
Includes index.
ISBN 978-1-107-01857-0 (hardback)
1. Submillimeter waves. 2. Electromagnetic waves. 3. Physics. 4. Terahertz spectroscopy. I. Title.
TK7877.L49   2012
539.2 – dc23   2012027118

ISBN 978-1-107-01857-0 Hardback

Cambridge University Press has no responsibility for the persistence or accuracy of
URLs for external or third-party internet websites referred to in this publication,
and does not guarantee that any content on such websites is, or will remain, accurate
or appropriate.

To my parents Frank and Julie

Contents

	Preface		*page* xi
1	**INTRODUCTION**		xiii
	1.1	Frequency. Hertz.	xiii
	1.2	Time. Second.	xiv
	1.3	Wavelength. Metres.	xv
	1.4	Wavenumber. Inverse metres.	xv
	1.5	Powers of ten	xvi
	1.6	Quantities, symbols, units, dimensions	xvi
	1.7	What lies ahead	xviii
	1.8	Summary	xix
Part I	**Basics**		1
2	**OSCILLATIONS**		3
	2.1	Describing oscillations	4
	2.2	Compact notation	11
	2.3	The time-bandwidth theorem	15
	2.4	The Fourier theorem	18
	2.5	Summary	23
	2.6	Table of symbols, Chapter 2	24
3	**COMBINING OSCILLATIONS**		25
	3.1	Oscillations in the same direction	26
	3.2	Oscillations in two dimensions	36
	3.3	Jones notation	48
	3.4	Summary	52
	3.5	Table of symbols, Chapter 3	53
4	**LIGHT**		54
	4.1	Photons	55
	4.2	Electromagnetic waves	56

	4.3	Describing waves	60
	4.4	Describing electromagnetic waves	62
	4.5	Polarisation	64
	4.6	Blackbody radiation	65
	4.7	Summary	72
	4.8	Table of symbols, Chapter 4	73
5	**MATTER**	**74**
	5.1	Wave mechanics	74
	5.2	The Schrödinger equation	75
	5.3	Free particle	77
	5.4	Infinitely high, square potential well	80
	5.5	Finite, square potential well	84
	5.6	Multiple, finite, square potential wells	86
	5.7	Oscillations of molecules	87
	5.8	Summary	96
	5.9	Table of symbols, Chapter 5	97
6	**INTERACTION OF LIGHT AND MATTER**	**98**
	6.1	A static electric field in a region of space	99
	6.2	Electromagnetic waves in matter	101
	6.3	Interaction of light and matter – harmonic oscillator model	104
	6.4	Electromagnetic waves in an insulator	112
	6.5	Electromagnetic waves in a conductor	113
	6.6	Electromagnetic waves at an interface	118
	6.7	Summary	122
	6.8	Table of symbols, Chapter 6	123

Part II Components   125

7	**SOURCES**	**127**
	7.1	Survey of sources of terahertz-frequency electromagnetic radiation	128
	7.2	Electromagnetic radiation from free charges – vacuum electronics	129
	7.3	Electromagnetic radiation from bound states – lasers	140
	7.4	Sources based on vacuum electronics	144
	7.5	Laser sources	146
	7.6	Sources driven by continuous lasers	149
	7.7	Sources driven by pulsed lasers	151

	7.8	Summary	160
	7.9	Table of symbols, Chapter 7	161
8	**OPTICS**		**162**
	8.1	Quasi-optics	163
	8.2	Mirrors	166
	8.3	Lenses	174
	8.4	Lightpipes and waveguides	180
	8.5	Prisms	180
	8.6	Attenuators	183
	8.7	Filters	183
	8.8	Windows	185
	8.9	Polarisers	186
	8.10	Summary	188
	8.11	Table of symbols, Chapter 8	189
9	**DETECTORS**		**190**
	9.1	Detector parameters	191
	9.2	Thermal detectors	195
	9.3	Electro-optic detectors	200
	9.4	Photoconductive detectors	201
	9.5	Summary	202
	9.6	Table of symbols, Chapter 9	203
Part III	**Applications**		**205**
10	**SPECTROSCOPY**		**207**
	10.1	Monochromatic sources	209
	10.2	Dispersive spectroscopy	210
	10.3	Interferometry	216
	10.4	Time-domain spectroscopy	223
	10.5	Summary	230
	10.6	Table of symbols, Chapter 10	231
11	**IMAGING**		**232**
	11.1	Passive imaging	234
	11.2	Raster imaging	239
	11.3	Focal plane array imaging	242
	11.4	Near-field imaging	242
	11.5	Tomography	246
	11.6	Summary	250
	11.7	Table of symbols, Chapter 11	251

Appendix A	**Prefixes**	253
Appendix B	**Mathematical symbols**	254
Appendix C	**Mathematics**	257
Appendix D	**Further reading**	262
	Glossary	263
	Index	271

Preface

This book aims to teach. It is directed to students. It is not directed to researchers.

Research literature is typically written in the third person. That is appropriate. In this book I will often write in the first person – there, I just did! – and address you – yes, you, the reader – in the second person. When I employ 'we' it is not as a royal plural, but as you and me together, author and reader, engaged in the common pursuit of applying ourselves to terahertz physics. I hope you are comfortable with, or will become comfortable with, this use of personal pronouns.

This book is a text book. So it contains text. But not unrelieved text. To give pedagogical structure, the text is divided into parts, into chapters, into sections and into subsections. But there is more, to accommodate different learning styles. Perhaps you are a visual learner; there are diagrams for you. Perhaps you are a native speaker of the language of mathematics; equations succinctly express much of the material. Perhaps you learn by example; there are examples sprinkled throughout. Perhaps you learn by doing; there are exercises aplenty. If you like reading dictionaries or telephone directories, you might like the glossary.

Apparatus to facilitate learning is integral to the book. Here is a fuller list of these features:

Assumed mathematics concepts are set out at the beginning of each chapter. The book is about physics, not mathematics, but mathematics is the language of physics, and a prerequisite to understanding much of what is said.
Learning goals near the beginning of each chapter identify the key pedagogical aims.
Examples concretely illustrate abstract concepts.
Exercises provide the opportunity to put new concepts into practice immediately. The exercises appear right where they are pertinent, rather than at the end of the section or at the end of the chapter. Doing the exercises is the best way to learn. I have learnt a lot through them.
Figures are used extensively and often drawn accurately to scale to provide additional information.
Key terms are cited at the end of each chapter, with a page reference to their main appearance.
Key equations are reiterated at the end of each chapter.
Tables of symbols appear at the end of each chapter. The abundance of concepts and the scarcity of symbols inevitably leads to the same symbols being used to

mean different things. Often the meaning is clear from context. Listing the symbols chapter by chapter reduces the likelihood of symbol confusion.

Glossary An extensive glossary is provided at the back of the book. Glossary terms appear in bold in the main part of the book.

Appendices set out common mathematical symbols used throughout the book, mathematical background material, information about physical quantities and further reading.

Index of many terms. Citations to figures are shown in bold in the index.

All the tools I have to hand – words, pictures, equations – I have employed to try to make my meaning clear. It is not bad that you encounter the same material expressed in words, in pictures, in equations. It is not an oversight. It is a deliberate strategy to assist you to learn. Repetition is the key to learning. Repetition is the key to learning. Whether prose, structure, pictures, equations, examples, exercises, notes or definitions speak most clearly to you, I hope you will find between the covers of this book a path to a greater understanding of terahertz physics.

'Where did that spring from?' is the despairing cry of the bamboozled student who can't fathom how a new topic builds on what has gone before. The sequencing of topics is critical to the pedagogy. I start with the basic and from there advance. For example, oscillations, involving one variable, come first. Waves, involving two variables, wait in the wings until Chapter 4. Purely mathematical concepts precede physical concepts. Physics is about energy and matter and even these are introduced separately, energy in Chapter 4, matter in Chapter 5. I have spent a lot of time thinking about the logical development of the subject so you don't have to give it a second thought.

Terahertz physics is not a completely distinct category of physics, as *nuclear physics* or *quantum physics* are. The idea of terahertz physics is similar to the idea of *high-energy physics*: it simply refers to the physics that obtains in a particular region. Hence terahertz physics does not refer to radically new phenomena, but to the application of the fundamental principles of physics – electromagnetism, mechanics, quantum theory – to a distinct subset of phenomena, namely, those that have characteristic rates of the order of a million million per second. So in studying this book, do not expect to learn fundamentally new physical principles, but expect to apply the physics you may already have in new circumstances. Perhaps you will even better understand some principles you already have a nodding acquaintance with, in seeing them in action in the terahertz region. If, after reading this book, you were to say, 'I never really understood (Fourier theory/dipole radiation/beats/...) before, but now I do', I would be delighted.

Thanks to Wendy and Warwick for spurring me on in different ways. Thanks to Joshua and Alaric for their comments. Thanks to Helen for endless love. Thanks to God, for from him and through him and for him are all things. To him be the glory forever! Amen.

R. A. Lewis
Wollongong
20–02–2012

1 INTRODUCTION

What is meant by frequency and why is it more fundamental than time?

The only mathematics you need to begin with is the ability to count and to perform the basic arithmetical operations of multiply and divide.

1.1 Frequency. Hertz.

> The unit of frequency is the hertz. Your heart rate is near to one hertz.

Let's start with the idea of **frequency**. (Rate means the same thing.) Frequency is a key concept and this book is built on it.

Take your pulse. Its frequency is about one hertz. This book is all about things that happen at a much greater rate, at about one terahertz. I will define a terahertz a little later, but for now it is enough to know that it is a rate. Let's begin with something closer to hand, the rate at which your heart pumps your blood through your veins, which you can feel for yourself at your neck or your wrist.

Your heart rate is near to one hertz. If you are resting it might be a tad less. If you get up and run, it might rise to two hertz or even three. You could check the change against a friend who is still at rest. You might find after running you counted twenty pulses for every ten of her pulses, whereas when you were sitting you had measured only twelve pulses for every ten of hers. It is easy to compare rates. No measuring instrument or equipment or apparatus of any kind is required. All you have to do is count.

Rather than test your heart rate against someone, you might calibrate it against something. Galileo is said to have compared his pulse rate to the rate at which a lamp, hanging on a long chain, swung back and forth in the Cathedral at Pisa. Say Galileo counted eight pulses as the chandelier went to then fro. Let's imagine the church bells rang twice as this was happening. Then the rate of Galileo's pulse was eight times the rate of the chandelier swing and the rate of the bells pealing was twice the rate of the chandelier swing. In such manner we can measure various rates. Not only is the order of the rates found – the chandelier the least, next the bells, then the pulse – but the numerical standings are also found – in the ratio one to two to eight. We do all this by counting. We do not need special instrumentation to measure rate.

We have seen that rates relative one to the other can be found out by counting. To go further, it is useful to have a benchmark rate to act as a standard and then to compare any other rate with it. My heart rate is most likely different from yours, so human heart rate would not be a well-defined standard. I might decide to ignore you and use my own heart rate as a standard, but I know my heart rate increases with exercise, so this would not be a good choice. We want as a standard something that is independent of a particular implementation and always has the same rate, as far as we can tell. How we can tell the rate is not changing is an interesting question, but we will answer it here by saying we assume or define the standard rate to be unchanging. Here is the standard of rate that will be used in this book:

> The rate of radiation corresponding to the transition between the two hyperfine levels of the ground state of the caesium 133 atom is 9 192 631 770 hertz.

This statement defines the **hertz**. The statement is consistent with the international system of units known as SI although not formally part of it. I don't expect you to follow the technical detail of the definition but you should appreciate two things. First, we are assuming that all caesium 133 atoms are the same – unlike human hearts, which differ from person to person. Second, we are assuming that a caesium 133 atom always radiates at the same rate – unlike human hearts, which pump at a greater rate during exercise than when at rest.

1.2   Time. Second.

> The unit of time is the second.

What is **time**? That is a good question. The answer is often not so good. 'Time is the ... ahh, time ... between two events'. Look in some dictionaries to see how self-referential and even tautological the definitions of time are.

How does one measure time? Use a clock. An instrument is needed to measure time. This is in contrast to rate, where we have seen that no device, only an ability to count, is needed.

In this book frequency is primary, time is secondary. Time is the inverse of frequency. I have defined the unit of rate, the hertz, already. Now, based on the hertz, I will define the unit of time, the **second**:

> The second is the inverse of the hertz.

By this I mean time [in seconds] multiplied by the rate [in hertz] amounts to one. So a heart rate of one hertz corresponds to a time between pulses of one second. A heart rate of two hertz corresponds to a time between pulses of half a second. As one quantity goes up, the other goes down; the product remains the same. This inverse relation may appear trivial but it has profound consequences. It would not be an exaggeration to say that much of this book turns on the inverse relation between frequency and time.

1.3 Wavelength. Metres.

> A wave has a particular length. It is called the wavelength. It is measured in metres.

I won't go into what is meant by a wave here (that is explained later) except to say that it has a characteristic length. This is the distance between two points on the wave that behave identically, for example, between two successive crests or between two successive troughs of an ocean wave. This distance is the **wavelength**.

Light has a wave nature. For light in empty space, the product of the wavelength and the frequency is constant. From this, we can define the unit of length, the **metre**:

> The metre is the wavelength, in a vacuum, of light of frequency 299 792 458 hertz.

This definition is consistent with the SI definition, albeit expressed differently.

A convenient multiple of the metre is often used when discussing waves. For example, millimetre waves refer to waves of wavelength about one thousandth of a metre. The field of terahertz is sometimes identified with millimetre or with sub-millimetre waves.

1.4 Wavenumber. Inverse metres.

> The wavenumber says how many waves there are in a given length.

How many waves there are in a given distance can be counted or numbered, and this is the origin of the word **wavenumber**.

A metric ruler is usually marked in millimetres and centimetres. There are typically nine strokes denoting the millimetres between two larger strokes marking successive centimetres, so over any portion of the ruler one centimetre in length there are ten strokes. The strokes are found at the rate of ten per centimetre. We say the linear frequency or spatial frequency of strokes is ten per centimetre, or ten inverse centimetres.

The SI unit for wavenumber is (number) per metre, or **inverse metre**. In spectroscopy – and there is some spectroscopy in this book – it has been traditional to count the number of waves per centimetre, rather than per metre. So you will often see the unit of inverse centimetre employed in specifying wavenumber. (Some people call the unit of inverse centimetre itself the wavenumber. I mention this only so that you are aware of it. I will avoid this usage in this book, however, as it opens the gate to confusion.) It may seem redundant, but for completeness let's define the unit for measuring wavenumbers:

> The inverse metre is the inverse of the metre.

Example 1.1 Heartbeats

A mouse has a heart rate of 10 hertz. The time between heartbeats in an elephant is 2 seconds. How much greater is the mouse's heart rate than the elephant's?

xvi INTRODUCTION

Solution
We convert the time information given for the elephant into a rate. Taking the inverse of 2 seconds, we arrive at 0.5 (the inverse of 2) hertz (the inverse of seconds). Thus the ratio of heart rates is 10/0.5 = 20. *Aside: The life span of many mammals is about one thousand million heartbeats. An elephant lives about 20 times as long as a mouse.*

1.5 Powers of ten

I have mentioned in passing 'millimetres' and 'centimetres'. I expect you already knew, or have gathered, a millimetre is one thousandth of a metre and a centimetre is one hundredth of a metre. The prefixes 'milli' and 'centi' are examples of a shorthand way to represent powers of ten. A full list of SI prefixes is given in Appendix A.

As far as this book is concerned, the most important power of ten is ten raised to the power of twelve, or 10^{12}. This is given the prefix 'tera'. This book is about the **terahertz**:

$$1 \text{ terahertz} \equiv 10^{12} \text{ hertz}.$$

This is the definition of the terahertz. I use the symbol '≡' to stand for 'is defined to be'. (It is stronger than '=', which means is 'equal to'. One terahertz does not just happen to be the same as 10^{12} hertz at the moment; it always is 10^{12} hertz.)

You can consult Appendix A to find many equivalent ways to express one terahertz. Here are some that are in common use:

$$1 \text{ terahertz} = 1\,000 \text{ gigahertz} = 1\,000\,000 \text{ megahertz}.$$

So a terahertz is a 'megamegahertz', meaning a million million hertz. That's a lot of hertz.

Exercise 1.1 Wavelength-wavenumber product
Show that the product of the wavelength, when expressed in micrometres, and the wavenumber, when expressed in inverse centimetres, is always 10 000. ∎

1.6 Quantities, symbols, units, dimensions

The only mathematics needed in this section – apart from the ability to count, to multiply and to divide – is basic algebra; that is, the ability to represent quantities by symbols.

We have met four quantities: frequency, time, wavelength and wavenumber. Rather than write these words out in full each time, it is easier to abbreviate each to a symbol. The simplest symbols are single characters. The characters I will use are set out in Table 1.1.

1.6 Quantities, symbols, units, dimensions

Table 1.1 Basic quantities and symbols

Quantity	Symbol
frequency	f
time	t
wavelength	λ
wavenumber	σ

The symbols are all shown in italic font. This is to better distinguish them from normal words. The symbols for frequency and time are rather straightforward in origin, being the first letters of those two words. The symbol for wavelength is the Greek letter lambda, the equivalent of the English letter 'l', which is associated with the first letter of 'length'. The symbol for wavenumber is the Greek letter sigma, the equivalent of the English letter 's'; the origin of the association is obscure.

The symbols save space, but at a cost. There are only a few different characters (compared to the huge number of words) and at different times the same character is used to represent different things. So t might be used to represent thickness and not time in another context. You need to take care with this.

When symbols represent physical properties, the symbols represent not only the numerical value of the quantity but also its unit. The symbols carry these two separate things. For example, f may represent '10 hertz'; both the quantity, here 10, and the unit, here hertz, are bound up in the single character f.

Apart from compactness, a second advantage of using symbols for quantities is that the symbols lend themselves easily to mathematical manipulation. This is known as 'quantity calculus'. So, rather than write 'inverse of the rate', I can write $1/f$, and calculate, for example, for a rate of 2 hertz,

$$\frac{1}{f} = \frac{1}{2\,\text{hertz}} = \frac{1}{2} \times \frac{1}{\text{hertz}} = 0.5\,\text{seconds}.$$

Notice that the units are bound up in the calculation. The answer, in this case, is in units of inverse hertz, or seconds.

Sometimes I will explicitly show the units that I am using. I will do this using square brackets, such as [in hertz]. For example,

$$\lambda\,[\text{in metres}] = \frac{1}{\sigma\,[\text{in inverse metres}]}.$$

Multiplying on both sides by the denominator,

$$\lambda\,[\text{in metres}] \times \sigma\,[\text{in inverse metres}] = 1\,[\text{unitless}],$$

where the notation [unitless] indicates a quantity without units, or a simple number. (Quantities like this, with no units, are called 'dimensionless quantities' or 'quantities of dimension one'.) I am most likely to show the units explicitly if I am using units that are not the usual SI units.

Just as we can use a compact notation to write quantities, we can use a compact notation to write units. The four key units and their abbreviations are set out in Table 1.2.

Notice these compact ways of writing the units are not in italic font. This distinguishes them from quantities. Still, there is a chance for confusion, for example if 'm'

Table 1.2 Basic units and abbreviations

Unit	Abbreviation
hertz	Hz
second	s
metre	m
inverse metre	m^{-1}

is used to represent the unit of metre and m has been chosen to represent the quantity of mass. Take care!

The unit 'hertz', when written out in full, begins with a lower-case 'h', but when abbreviated as 'Hz' begins with an upper-case 'H'.

The power-of-ten prefixes used with units also have abbreviated forms. For example, the prefix 'mega', meaning one million, is abbreviated as 'M'. A full list is given in Appendix A.

In stating numerical values given with units, a space should be left between the two. So, write '5 THz', not '5THz'.

Using the abbreviations for prefixes and units, we may more compactly write

$$1\,\text{THz} = 1\,000\,\text{GHz} = 1\,000\,000\,\text{MHz}.$$

Apart from the physical quantity itself and its units, and how to express these in compact, symbolic forms, it is useful to be aware of dimensions. For example, wavelength has the dimension of length, which we indicate by the sans serif font L. Wavenumber has the dimension of inverse length (L^{-1}). We may assign to frequency the dimension F. Then time will have the dimension F^{-1}.

1.7 What lies ahead

I hope you now have a feel for the quantities of frequency and wavelength, and their inverses, time and wavenumber. Frequency is associated with oscillations. Wavelength is associated with waves. Much of the book is concerned with these two topics.

Now is a good time to give an outline of the rest of the book.

Part I Basics Part I lays the foundation for the rest of the book. Oscillations are the subject of Chapters 2 and 3. Chapter 2 provides a framework for dealing with oscillations. Chapter 3 explains how oscillations are combined. These foundational chapters are largely mathematical in character. On to physics. Physics is about energy and matter and their interaction. Chapter 4 introduces energy in its purest form – light. Chapter 5 introduces matter and the phenomena that occur in matter at terahertz frequencies. Chapter 6 deals with the interaction of light and matter. Chapter 3 is built on Chapter 2; Chapter 4 is built on both of these, as is Chapter 5; Chapter 6 is built on Chapters 4 and 5.

Part II Components Part II describes the building blocks that make up practical terahertz systems. We start with the sources of terahertz-frequency electromagnetic radiation in Chapter 7. As the terahertz radiation propagates, it can be manipulated by various optics, the subject of Chapter 8. Finally, Chapter 9 deals with the radiation being detected at its journey's end. These three chapters may be read independently of each other, but all depend on the foundation of Part I.

Part III Applications Part III describes two important areas of application of terahertz technology. Chapter 10 describes spectroscopy. Chapter 11 describes imaging. These chapters may be read independently of each other, but are each based on the terahertz components described in Part II.

From the way I have set this out, you may think of a building going up and being used: Part I is about the foundations that underpin the whole structure; Part II is about the bricks and mortar that together constitute the building; Part III is about the uses to which the finished premises are finally put.

Or think of a tree. Part I is like the roots, often unseen, but sustaining the rest. Part II is like the leaves and branches, what we take as the tree. Part III is like the fruit, the product of the whole.

1.8 Summary

We have met four physical quantities: frequency, time, wavelength and wavenumber. Each may be represented by a simple symbol. Each has a unit and a dimension. (We have also met the abbreviated units and prefixes for units.) Table 1.3 summarises these.

Table 1.3 Summary of basic physical quantities

Quantity		Unit		Dimension	
frequency	f	hertz	Hz	frequency	F
time	t	second	s	inverse frequency	F^{-1}
wavelength	λ	metre	m	length	L
wavenumber	σ	inverse metre	m^{-1}	inverse length	L^{-1}

This book is all about things that happen at a rate of about one terahertz:

$$1\,\text{THz} \equiv 10^{12}\,\text{Hz} = 1\,000\,\text{GHz} = 1\,000\,000\,\text{MHz}.$$

Part I
Basics

2 OSCILLATIONS

You don't know anything about trigonometry? That's OK, I'll teach you all you need to know as we go along. I will introduce summation notation, but explain it. If you know how to integrate, that might be an asset, but it is not strictly necessary as I will give you a visual description of what is involved. This should allow you to appreciate the meaning of the equations even if you do not have a full grasp of the apparatus of integration.

In this chapter we meet *oscillations*. We will look at the general way to describe any oscillation using mathematics.

To describe an oscillation in mathematical terms, we identify three key properties: how rapid it is, how large it is and when it starts. These three properties are more formally defined as *frequency*, *amplitude* and *initial phase*. We can express an oscillation mathematically by using a trigonometric function such as cosine or sine or by using a compact exponential notation involving complex numbers.

The *time-bandwidth theorem* appears over and over again in terahertz physics. It says that the product of the duration of a pulse (time) and the range of frequencies encompassed in the pulse (bandwidth) has a minimum value. Looked at in one way, if we have a short pulse, the pulse must involve a large range of frequencies. Looked at in another way, a well-defined frequency implies a very long pulse. This profound yet simple concept has wide-ranging ramifications in the production, detection and application of terahertz-frequency electromagnetic radiation.

Fourier methods play a huge role in modern physics and engineering. Here is the basic concept in a nutshell: an arbitrary oscillation can be constructed from fundamental oscillations. The Fourier method can be applied coming or going. In *Fourier synthesis*, one or more fundamental oscillations are added together to produce the final desired oscillation (which can be more or less anything you want). In *Fourier analysis*, a given, and possibly quite complicated, oscillation is broken up into its constituent fundamental oscillations. Both Fourier synthesis and Fourier analysis find many applications in terahertz physics. A good appreciation of their power is essential; an appreciation of their beauty is optional.

Learning goals

There are three important things I want you to know by the time you finish this chapter:

- how to describe oscillations in mathematical terms,

- the meaning of the time-bandwidth theorem,
- the importance of Fourier methods.

2.1 Describing oscillations

Let us start with a very simple example. About the simplest oscillation I can think of is something turning on and off. Something like a flashing light. The sequence on, off, on, off, on, off, on, repeats over and over.

Figure 2.1 represents graphically the very simple oscillation of a light switching on and off.

Let us observe first that the phenomenon I have chosen to describe is **periodic**: the same thing keeps happening over and over again. The prime characteristic of this phenomenon is its **frequency** (Figure 2.1a). The frequency is the number of oscillations per unit time. In this example the frequency happens to be 5 oscillations per minute. This frequency corresponds to the light being on for 6 seconds, then off for 6 seconds, and so on (Figure 2.1b).

Let us now elaborate on this example a little. Let us now consider measuring the amount of light more accurately, rather than just saying the light is 'on' or 'off'. Let's say we found a meter and measured the light to produce an illuminance of 100 lux. Then we could represent the oscillation more fully as in Figure 2.1c and Figure 2.1d, where a label is added to each vertical axis to denote the size of the oscillation. The size of the oscillation is its **amplitude**. More precisely, the amplitude is measured as the swing above and below the average value of the oscillation. In this example, the average value is 50 lux, and the illuminance swings 50 lux above and below this. So the amplitude is formally defined as 50 lux in this example (Figure 2.1d), not 100 lux. The amplitude has units that depend on the quantity being measured. For example, the temperature in a room might be oscillating and the amplitude would then be measured in temperature units, such as Celsius degrees. In the case of terahertz phenomena we are often interested in the amplitude of an electric field, and this is measured in the units of volts per metre.

Now look at Figure 2.1e and Figure 2.1f, where a different oscillation is shown. This second oscillation corresponds to a second light being turned off and on repeatedly. The oscillation in Figure 2.1e/f differs from the oscillation in Figure 2.1c/d in three key respects. Pay attention, for these are the three fundamental characteristics of an oscillation. First, the two oscillations differ in *frequency*. We have seen already that the first oscillation has a rate or frequency of 5 per minute. The second oscillation has a rate of 3 per minute (Figure 2.1e). This smaller rate corresponds to a greater time for the second lamp being turned on and off; it is off for 10 seconds, on for 10 seconds, and so on (Figure 2.1f). Second, the two oscillations differ in size, in *amplitude*. We have seen already that the first oscillation has an amplitude of 50 lux. The second oscillation has an amplitude of 150 lux. Third, the two oscillations *start at different points*. Taking the far left of Figure 2.1d and of Figure 2.1f to be the beginning of our measurement,

2.1 Describing oscillations

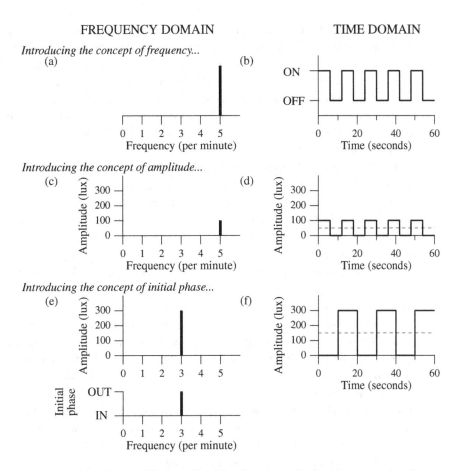

Figure 2.1 Describing oscillations. The figure introduces the three key properties of an oscillation: *frequency*, *amplitude* and *initial phase*. These properties are illustrated in both the frequency domain (left-hand side of the figure) and in the time domain (right-hand side of the figure). Panel (a) represents an occurrence with a *frequency* of 5 times per minute, such as (b) a light being repeatedly turned on for 6 seconds then off for 6 seconds. Panels (c) and (d) now specify the *amplitude* of the phenomena – the light provides an illuminance of 100 lux. (A lamp in your house might give this much illumination.) Panels (e) and (f) represent another light, giving an illuminance of 300 lux, first being turned off for 10 seconds, then on for 10 seconds, and so on. This oscillation differs from the previous oscillation in all three key characteristics of frequency, amplitude, and *initial phase*.

we see the first oscillation starts at the value of 100 lux, whereas the second starts at the value of zero. Had we only one oscillation, we may not have noticed this subtlety, but with two oscillations, we see they are out of step at the outset. The **initial phase** tells us where we are along the complete cycle of an oscillation at a reference time. The reference time usually chosen is the time when the measurement begins. Seeing the two oscillations helps us recognise there is a difference in initial phase, but we do not need two oscillations; we can define initial phase for a single oscillation by noting how far along a complete cycle we are at the reference time. I have been assuming that the first oscillation has zero phase, meaning that I count the oscillation as beginning when the light is switched on. I have not explicitly shown this zero initial phase in Figure 2.1c. The second light is halfway through its cycle when we start the measurement (Figure 2.1f); we have to wait 10 seconds, or half a cycle, before it is switched on. I count this as being completely out of phase. The phase of the second oscillation is indicated on a second vertical axis in Figure 2.1e.

We can choose to describe oscillations in the **frequency domain** (the left-hand side of Figure 2.1). In the frequency domain, we see explicitly what different frequencies are present, their relative amplitudes and initial phases. Alternately, we may describe oscillations in the **time domain** (the right-hand side of Figure 2.1). Oscillations shown in the frequency domain and the time domain amount to the same thing; equivalent information is in each representation. However, depending on our purpose, working in one or other domain may prove to be more convenient. This book is about phenomena characterised by particular (terahertz) frequencies, so prepare to spend some time in the (terahertz) frequency domain.

Let us now turn to a smoother example. A common form of periodic motion is rotation. We will start with the simplest version of rotation – motion on the simplest curve (the circle) and at the simplest rate (evenly). Such motion is called **uniform circular motion**. I expect you have studied uniform circular motion previously but it doesn't matter if you haven't. To a reasonable approximation, uniform circular motion may be used to describe the motion of the earth around the sun, or the motion of an electron around a proton in a hydrogen atom. Uniform circular motion is easy to grasp. In Figure 2.2a I show uniform circular motion by representing a particle moving in – you guessed it – a circle, and at – wait for it – a steady speed.

The situation is very simple, as described first in words (uniform circular motion) then described in a picture (Figure 2.2a). We may also describe this situation very simply using a third language, mathematics. Let's call the angle through which the particle has moved around the circle theta, θ. The first symbol I think of when I think of an angle is θ, so that is what I use here. Throughout the book, θ is employed as a general-purpose angle. We will measure the angle in radians. If you have not previously met the radian, abbreviated *rad*, all you need to know is it is used to measure angles around a circle, and moving once around the circle amounts to 2π radians (Figure 2.2b). More details on angles and their measurement can be found in Appendix C. Of course, as the particle is moving, the angle is always changing. But as the particle is moving uniformly, the angle is changing at a steady rate. Let's call this steady rate of angular change ω. Assuming we start with θ being zero at time zero then at time t

$$\theta = \omega t. \tag{2.1}$$

This relationship is illustrated in Figure 2.2c. For example, if the rate of change of angle θ is one twelfth of a circle per second, or $2\pi/12 = \pi/6$ radians per second, after 1, 2 and

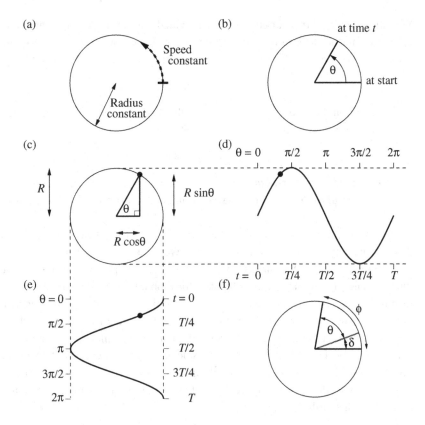

Figure 2.2 Uniform circular motion. (a) Motion at a steady rate around a circular path, in other words, uniform circular motion. (b) In a time interval t, motion at the steady angular rate of ω results in an angle of $\theta = \omega t$ being traversed. Here t is measured in seconds, ω is in radians per second and θ is in radians. The time taken to complete one lap of the circle is the period, T, and is measured in seconds. The period is directly related to the number of laps per second, or frequency, f, by $f = 1/T$. Frequency has units of hertz. The idea of laps per second may be expressed in terms of radians per second; there are 2π radians in one lap. The angular frequency is then $\omega = 2\pi/T = 2\pi f$. (c) Defining the trigonometric functions cosine and sine. (d, e) Uniform circular motion projected onto two perpendicular axes. (f) If the motion does not start at the usual origin, this is accommodated by an initial phase δ. The phase, ϕ, takes into account both the motion described previously (θ) and the initial phase (δ); $\phi = \theta + \delta$.

3 seconds, the particle will have moved one twelfth, then two twelfths (one sixth) and finally three twelfths (one quarter) of the way around the circle, Figure 2.2c.

Once the angle changes by a full circle, or $\theta = 2\pi$, the motion starts again. Putting $\theta = 2\pi$ into Equation (2.1), the time taken for one complete oscillation is $2\pi/\omega$. Since the frequency is the inverse of the time taken for one complete oscillation, $f = \omega/2\pi$. Making ω the subject of the formula,

$$\omega = 2\pi f. \tag{2.2}$$

We call ω the **angular frequency**. (Earlier I called this the steady rate of angular change; it amounts to the same thing.) Rather than express θ in terms of angular frequency we can express it in terms of frequency by substituting Equation (2.2) into Equation (2.1):

$$\theta = 2\pi f t. \tag{2.3}$$

By definition, being always on the circle, the distance from the centre of the circle is always the same. We will call this distance R.

$$R = \text{constant}. \tag{2.4}$$

The two-dimensional (or *plane*) motion about a centre (or *pole*) is conveniently expressed in the **plane-polar coordinates**, R and θ, as introduced in Equations (2.4) and (2.1).

In many circumstances, we like to have coordinates straight and true: not the mongrel distance and angle of the plane-polar system. We want to work in Cartesian **rectangular coordinates**, x and y. So our task is now to express the position of the particle at any time in x and y coordinates. Let us break up this job into two parts, first the x coordinate, then the y. To begin with (in other words, when time $t = 0$), the particle is actually on the x axis, and a distance R from the centre of the circle (the origin of the x-y axis system). So the x component is R. After the particle has moved through $\pi/3$ radians, which happens after 2 seconds in our example, the x component is $R/2$. (You could take my word for it, or measure it.) We could do the same for every angle – write down the projection onto the x axis (Figure 2.2e). Doing this produces the cosine function. In mathematical shorthand,

$$x = R\cos\theta. \tag{2.5}$$

If you have never met the cosine function before, note that this equation defines it. In words, for a given angle of rotation anticlockwise around a circle starting at the x axis, the cosine function gives the projection of the position onto the x axis (as a proportion of the radius). In a similar way, we can write the y component as

$$y = R\sin\theta. \tag{2.6}$$

In words, the sine function gives the y projection of the particle position as a proportion of the radius (Figure 2.2d). More detail about the cosine and sine functions is given in Appendix C.

If we want to show the time dependence explicitly, we may use Equation (2.1) to write

$$x = R\cos\omega t \quad \text{and} \quad y = R\sin\omega t, \tag{2.7}$$

or Equation (2.3) to write

$$x = R\cos 2\pi f t \quad \text{and} \quad y = R\sin 2\pi f t. \tag{2.8}$$

If we are only interested in the projection along a particular axis, we can use one (or other) of these two equations. Some prefer the cosine expression, and refer to cosinusoidal oscillations. Some prefer the sine expression, and refer to sinusoidal oscillations.

2.1 Describing oscillations

It really amounts to the same thing, as the cosine function and the sine function differ only by a rotation of one-quarter of a circle, in other words, by $\pi/2$ radians (Appendix C):

$$\cos\theta = \sin(\theta + \pi/2). \tag{2.9}$$

Thus, we may say cosine and sine differ only in initial phase. This is illustrated in Figure 2.2d/e. If we count the oscillation as beginning with the value 1 and decreasing from there, the cosine function is in phase and the sine function is one-quarter cycle, or $\pi/2$ radians, out of phase.

The discussion about the similarity between cosine and sine, apart from the starting point, brings us to the relatively minor matter of dealing with motion that does not start on the x axis at $t = 0$. We may regard this as either an offset in time or an offset in angle. (An example is the alternate starting point in Figure 2.2f.) The offset angle, δ, is how we will usually express initial phase from now on. This offset in angle corresponds to an offset in time, which we will denote by t_δ. The relation between the two quantities is

$$\delta = \omega t_\delta. \tag{2.10}$$

(This may be seen to follow from Equation (2.1).) The x value of an oscillation of initial phase δ (such as the oscillation shown in Figure 2.2f measured from the alternate starting point) may be represented by

$$x = R\cos(\omega t + \delta) = R\cos(2\pi f t + \delta). \tag{2.11}$$

The argument of a trigonometric function is called the **phase** and denoted ϕ. Note that the phase, ϕ, is distinct from the initial phase, δ. Here the phase is the argument of the cosine function,

$$\phi = \omega t + \delta = 2\pi f t + \delta. \tag{2.12}$$

Using this expression for phase we can write Equation (2.11) as

$$x = R\cos\phi, \tag{2.13}$$

putting it into exactly the same form as Equation (2.5).

So far, the physical quantity we have been focussing on has been the position in space. We can extend the idea of oscillations to other quantities, such as temperature, or pressure, or electric field. Using A to represent a general quantity and A_0 the amplitude of that quantity, we can write a **harmonic** oscillation in general as

$$A = A_0 \cos(\omega t + \delta) = A_0 \cos(2\pi f t + \delta). \tag{2.14}$$

Note that A and A_0 have the same dimensions so that the equation balances.

To sum up, the three main characteristics of an oscillation are its *frequency*, its *amplitude* and its *initial phase*. In mathematical terms, beginning from the idea of motion in a circle, we may write these three key parameters for circular, or harmonic oscillations as

Frequency	f	hertz	(2.15)
Amplitude	A_0	(units pertinent to the oscillation)	(2.16)
Initial phase	δ	radians	(2.17)

A general mathematical form to represent any harmonic oscillation is then:

$$A = A_0 \cos(2\pi f t + \delta) = A_0 \cos \phi. \quad (2.18)$$

The constants are A_0, f and δ, the three characteristics of the oscillation; the variable is the time, t; these four quantities are related to the size of the oscillation via the trigonometric function cosine. The phase, ϕ, incorporates the change with time, $2\pi f t$, and the initial phase, δ.

Example 2.1 The unit diamond

(This example assumes a knowledge of trigonometry. You can skip it if your trigonometric knowledge is weak or non-existent.) The definitions of cosine and sine were introduced in Figure 2.2 based on motion in a circular path. In a similar fashion, other functions may be defined based on motion along other paths. Consider motion along a *unit diamond* (Figure 2.3a); that is, a set of four straight lines running from (1, 0) to (0, 1), to (−1, 0), to (0, −1) and back to (1, 0). Give an expression for the projection of a position on the unit diamond onto the y axis as a function of angle θ (measured anticlockwise from the x axis). Also give an expression for the projection onto the x axis.

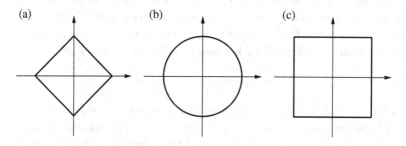

Figure 2.3 (a) The unit diamond. (b) The unit circle. (c) The unit square. The figures cross the axes at a unit distance from the origin.

Solution
Let us consider the first quadrant ($x > 0, y > 0$). From Figure 2.3a we see that x and y on the unit diamond are related by $x + y = 1$. Also, for any pair of x, y the definition of the tangent is $\tan\theta = y/x$. We may eliminate x from these two equations to obtain y. This is exactly what we are seeking, the projection of a point on the unit diamond onto the y axis. In analogy with the definition of sine, we write this as

$$\diamond \sin\theta = \frac{\tan\theta}{1 + \tan\theta}. \tag{2.19}$$

Likewise, eliminating y from the first two equations, we obtain

$$\diamond \cos\theta = \frac{1}{1 + \tan\theta}. \tag{2.20}$$

If a particle is moving at a steady speed around the straight edges of the unit diamond the plot of y (or of x) against time gives a triangular or sawtooth pattern. If the particle is moving at a steady angular speed, that is, if θ changes uniformly with time (or, equivalently, if ω is constant), then the plot of y against time is given by Equation (2.19) and the plot of x against time is given by Equation (2.20). In circular motion, the linear speed and the angular speed are directly proportional. This is not so for motion on the diamond.

Exercise 2.1 The unit circle
(This exercise assumes little knowledge of trigonometry. You should attempt it if your trigonometric knowledge is weak or non-existent.) The definitions of cosine and sine were introduced in Figure 2.2, based on steady motion in a circular path of radius R. They may be more simply defined without reference to R in relation to the unit circle, Figure 2.3b.
 (a) Define cosine and sine in relation to the unit circle.
 (b) Use Pythagoras' theorem to write down a relation between cosine and sine.
 (c) Invert Equations (2.5) and (2.6) to obtain R and θ in terms of x and y. ∎

Exercise 2.2 The unit square
(This exercise assumes good knowledge of trigonometry.) Consider motion along a *unit square* (Figure 2.3c). Give expressions for the projection of a position on the unit square onto the y axis as a function of angle θ ($\square\sin\theta$) and an expression for the projection onto the x axis ($\square\cos\theta$). ∎

2.2 Compact notation

An alternative way to write Equation (2.18) is

$$A = A_0 \exp[(2\pi ft + \delta)i]. \tag{2.21}$$

On the face of it, this is more complicated; it involves the exponential function rather than the cosine function and in addition involves the imaginary number i. But as we shall see, there are many advantages to using this notation, especially when we wish to multiply or divide oscillations. To work our way up to this equation, I will first give an account of complex numbers, then relate them to the exponential function.

Thus far we have been dealing with **real numbers**, such as x and y. Now I introduce the prototypical **imaginary number**, the square root of negative one, denoted i:

$$i = \sqrt{-1} \qquad i^2 = -1. \tag{2.22}$$

Multiplying a real number by an imaginary number gives an imaginary number; for example, iy is an imaginary number. Associating a real number, such as x, with an imaginary number, such as iy, results in a **complex number**. I will use a tilde (~) over a number to indicate it is complex. Let us call the combination of x and iy the complex number $\tilde{z}$:

$$\tilde{z} = x + iy. \tag{2.23}$$

An alternative notation for a complex number is

$$\tilde{z} = \text{Re}(\tilde{z}) + i\text{Im}(\tilde{z}), \tag{2.24}$$

where the real part of $\tilde{z}$ is

$$\text{Re}(\tilde{z}) = x \tag{2.25}$$

and the imaginary part of $\tilde{z}$ is

$$\text{Im}(\tilde{z}) = y. \tag{2.26}$$

Alternative notations again are to write the real part of $\tilde{z}$ as $\text{Re}(\tilde{z}) = z_1$ or $\text{Re}(\tilde{z}) = z'$ and likewise to write $\text{Im}(\tilde{z}) = z_2$ or $\text{Im}(\tilde{z}) = z''$. Like x and y, it is understood z_1, z_2, z' and z'' are real.

If $y = 0$, we say $\tilde{z}$ is **purely real**; if $x = 0$, we say $\tilde{z}$ is **purely imaginary** (unless $x = y = 0$, when we say nothing).

Complex numbers can be represented as points on the **complex plane**; see Figure 2.4. The real component is represented on the real axis and the imaginary component is represented on the imaginary axis. The coordinates (x, y) locate the complex number $\tilde{z}$. The distance of the point (x, y) from the origin gives the **modulus** or **absolute value** of $\tilde{z}$, denoted either mod $\tilde{z}$, $|\tilde{z}|$ or r:

$$\text{mod } \tilde{z} = |\tilde{z}| = r = \sqrt{x^2 + y^2}. \tag{2.27}$$

The angle the line joining the origin and point makes with the x axis gives the **argument** (or the phase) of $\tilde{z}$, denoted θ. (I am using θ to denote a general angle. It may take on the value given in Equation (2.1), or it may be something else.)

$$\arg \tilde{z} = \theta = \arctan \frac{y}{x}. \tag{2.28}$$

Using the plane-polar coordinates r and θ rather than the Cartesian coordinates x and y

2.2 Compact notation

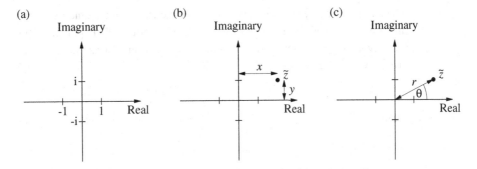

Figure 2.4 (a) The complex plane. The real axis is horizontal. The imaginary axis is vertical. (b) A complex number $\tilde{z}$ is represented by the point (x, y). Here x is the distance along the real axis and y is the distance along the imaginary axis. We write this as $\tilde{z} = x + iy$. The example shown is $\tilde{z} = 2 + i$. (c) A complex number $\tilde{z}$ is represented by a point. The distance from the origin to the point, r, is the *absolute value* or *modulus* of $\tilde{z}$: $r = |\tilde{z}| = \text{mod } \tilde{z}$. The line joining the origin and $\tilde{z}$ makes an angle with the x axis of θ. This is the *argument* of $\tilde{z}$: $\theta = \arg \tilde{z}$. We write $\tilde{z} = re^{i\theta}$.

provides an alternative way to express the real and imaginary components:

$$\text{Re}(\tilde{z}) = r \cos \theta, \tag{2.29}$$

$$\text{Im}(\tilde{z}) = r \sin \theta. \tag{2.30}$$

The **complex conjugate** of $\tilde{z}$ is formed by reversing the sign of the imaginary component. It is denoted $\tilde{z}*$. Following Equations (2.25) and (2.26),

$$\tilde{z}* = x - iy = \text{Re}(\tilde{z}) - i\text{Im}(\tilde{z}). \tag{2.31}$$

Adding and subtracting the complex conjugate to and from the original complex number gives

$$\tilde{z} + \tilde{z}* = 2x = 2\text{Re}(\tilde{z}) \tag{2.32}$$

and

$$\tilde{z} - \tilde{z}* = 2iy = 2i\text{Im}(\tilde{z}). \tag{2.33}$$

Inverting these equations gives

$$x = \text{Re}(\tilde{z}) = \frac{\tilde{z} + \tilde{z}*}{2} \tag{2.34}$$

and

$$y = \text{Im}(\tilde{z}) = \frac{\tilde{z} - \tilde{z}*}{2i}. \tag{2.35}$$

Multiplying a complex number with its complex conjugate yields the square of the modulus:

$$\tilde{z}\tilde{z}* = (x + iy)(x - iy) = x^2 - ixy + iyx - i^2y^2 = x^2 + y^2 = r^2. \tag{2.36}$$

Having met complex numbers, let's meet the exponential function. The exponential function of a variable v is written either as $\exp[v]$ or e^v. The exponential function is defined by a unique property: differentiating the function returns the function. In symbols, $d(e^v)/dv = e^v$. Or, in two steps,

$$u = \exp[v] = e^v \qquad (2.37)$$

means

$$\frac{du}{dv} = u. \qquad (2.38)$$

Having defined the exponential function, let's substitute for u the complex number of modulus 1 and argument θ:

$$\tilde{w} = \cos\theta + i\sin\theta. \qquad (2.39)$$

Differentiating with respect to θ,

$$\frac{d\tilde{w}}{d\theta} = -\sin\theta + i\cos\theta. \qquad (2.40)$$

(Details of differentiating the cosine and sine functions are given in Appendix C.) The right-hand sides of Equations (2.39) and (2.40) are intriguingly similar. In fact, if we divide Equation (2.40) through by i (noting that $1/i = -i$), we find

$$\frac{d\tilde{w}}{d[i\theta]} = \cos\theta + i\sin\theta = \tilde{w}, \qquad (2.41)$$

where the last equality follows from Equation (2.39). Equation (2.41) is precisely of the form of Equation (2.38). So we can apply Equation (2.37) making the identifications $u = \tilde{w}$ and $v = i\theta$. Thus the curtain rises on a profound mathematical result, **Euler's theorem**:

$$e^{i\theta} = \cos\theta + i\sin\theta. \qquad (2.42)$$

Let's apply Euler's theorem. To extend from a unit complex number $\tilde{w}$ to a complex number of modulus r, we simply multiply both sides of the equation by the magnitude r:

$$\tilde{z} = re^{i\theta} = r\cos\theta + ir\sin\theta. \qquad (2.43)$$

How does this relate to oscillations? Well, if you look back to Equation (2.18) you will notice some similarity with Equation (2.43). In fact, if we replace r with A_0 and θ with ϕ in Equation (2.43), we have

$$\tilde{z} = A_0 e^{i\phi} = A_0 \cos\phi + iA_0 \sin\phi. \qquad (2.44)$$

The real part, $A_0 \cos\phi$, is precisely the meaning of A from Equation (2.18).

$$A = \text{Re}(A_0 e^{i\phi}) = A_0 \cos\phi. \qquad (2.45)$$

The imaginary part, $iA_0 \sin\phi$, is a sort of add-on extra we don't need.

You may well ask, why go to all this trouble to introduce complex numbers, to introduce exponential notation, to end up with an equation, half of which is discarded? Good question; as you will see, it does turn out to be worth the trouble and in fact provides a very elegant way of manipulating oscillations.

Now I come to something I am a little embarrassed about. It is a bit sloppy, but it proves to be *so* convenient to work in the complex plane that many people – myself included – often simply write

$$A = A_0 e^{i\phi}. \tag{2.46}$$

You can see that Equation (2.46) is inconsistent with Equation (2.45). For a start, on the left and the right of Equation (2.45) are explicitly real numbers; the right-hand side of Equation (2.46) is a complex number, whereas the left-hand side is not indicated as being complex and, we have been assuming, is real.

When you come across an equation like (2.46), you should interpret it as follows. The oscillation is real, and is given by $A_0 \cos\phi$. We add to this the imaginary term $iA_0 \sin\phi$ and write the combination as the complex number $A_0 e^{i\phi}$. After some manipulations in the complex plane, we obtain an answer, still a complex number. Finally, we extract the real part from the answer to obtain the required oscillation.

Exercise 2.3 Complex numbers
For the complex number $1 + i$ write down (a) the real part, (b) the imaginary part, (c) the modulus, (d) the argument, (e) the complex conjugate. ∎

Exercise 2.4 Euler's theorem
Use Euler's theorem to determine the value of (a) e^0, (b) $e^{i\pi/2}$, (c) $e^{i\pi}$, (d) $e^{2\pi i}$. ∎

Exercise 2.5 de Moivre's theorem
Starting from Euler's theorem, derive *de Moivre's theorem*: $(\cos\alpha + i\sin\alpha)^n = \cos n\alpha + i\sin n\alpha$. ∎

2.3 The time-bandwidth theorem

Imagine you visit the doctor for a checkup. (It's OK, you are not ill, it's just a checkup.) The first thing the doctor does is measure your heart rate. Let's say she counts the number of heartbeats in a minute and gets 60. That is pretty normal. If she'd stopped a little sooner she might have only counted 59, or stopped a little later she might have counted 61, so let's say the result is 60±1 beats per minute (bpm). For the purposes of this example, I'll assume your heartbeat is perfectly regular, so the inaccuracy in the result does not come from variation in your pulse rate. Rather, the inaccuracy in the result comes from exactly when the doctor started and stopped timing.

Now, let's say it's a busy day in the surgery, and she only counted for half a minute. We would expect she would measure 30, or perhaps 29, or perhaps perhaps 31 beats in the half minute. These three possibilities translate to 60 or 58 or 62 bpm. In other words, the result is 60±2 bpm. Counting for only 10 seconds she may have recorded 9, 10 or 11 beats, amounting to 60±6 bpm. Were she to count for only half a second, sometimes she would miss a heartbeat, and record that your heart had stopped; other times, catch a heartbeat, and convert that to the rather rapid 120 bpm. This last result can be written as 60±60 bpm. It should be clear that the shorter the time interval over which the measurement is taken, the greater the range of uncertainty in the measured frequency. In fact, as in this example, the product of the time interval and the frequency range remains constant. Here a 60-second timing interval led to a frequency range of 1 bpm, a 30-second interval to 2 bpm and a 10-second interval to 6 bpm. The product of the time interval and the frequency range is constant. This is the heart – pun intended – of the time-bandwidth theorem.

The time-bandwidth theorem states that the product of the time interval and the frequency range is never less than one:

$$\Delta f \Delta t \geq 1. \tag{2.47}$$

The time-bandwidth theorem may also be written in terms of angular frequency as

$$\Delta \omega \Delta t \geq 2\pi. \tag{2.48}$$

In the case of a single, constant frequency source, the time-bandwidth theorem tells us that the precision with which this frequency may be measured increases as the time of measurement increases. This was demonstrated in the example at the doctor's surgery. Your heartbeat was presumed to be at a single, unvarying frequency. To make any sensible measurement of frequency at all, you need to measure for a time at least equal to the inverse of the frequency. In this example, to make sure you catch a heartbeat, you need to measure for a second. The longer you measure, the more precisely you determine the frequency.

The time-bandwidth theorem works the other way as well. By that I mean, as a phenomenon is restricted to a shorter time, the frequencies involved are spread over a greater range. This application of the time-bandwidth theorem is the basis of the generation of terahertz radiation using short pulses of (usually near-infrared) light. The shorter the pulses of light used in generating the terahertz, the broader the bandwidth of the radiation produced.

The time-bandwidth theorem is illustrated in Figure 2.5.

Example 2.2 Picosecond pulses

(This example is very simple.) One way of generating terahertz-frequency radiation is by using short pulses. Often the method is described as needing picosecond or sub-picosecond pulses. What is the bandwidth that might be generated using a picosecond pulse?

Solution

In solving this problem we make use of the time-bandwidth theorem, Equation (2.47). We are told the duration of the pulse is 1 picosecond, in other words, 10^{-12} seconds. Thus $\Delta t = 10^{-12}$ s. Substituting this into Equation (2.47) and solving for Δf gives

2.3 The time-bandwidth theorem

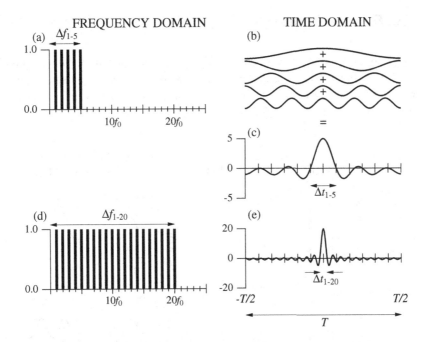

Figure 2.5 The time-bandwidth theorem. (a) Frequency-domain representation of oscillations at five different frequencies, $f_0, 2f_0, 3f_0, 4f_0$ and $5f_0$. (b) Time-domain representation of the five oscillations. (The amplitude of each is 1; they have been offset in the vertical direction for clarity.) (c) Result of adding together the five oscillations. Shown below is the addition of twenty oscillations of different frequencies in (d) the frequency domain and (e) the time domain. As the frequency range (or bandwidth) increases fourfold, panels (a) and (d), the width in the time domain decreases fourfold, panels (c) and (e). The product of the time and the bandwidth remains constant. Although this figure only shows a simple combination of oscillations (the oscillations all have the same amplitude and all have the same initial phase), the time-bandwidth theorem holds more generally.

$\Delta f = 10^{12}$ inverse seconds $= 10^{12}$ Hz. Thus the generated radiation is expected to have a bandwidth of 1 THz.

Exercise 2.6 Femtosecond pulses
Terahertz-frequency radiation is generated using pulses of duration 12 fs. What is the bandwidth generated? ∎

Exercise 2.7 Gaussian pulses
(This exercise assumes good knowledge of Fourier transforms and of Gaussian pulses.) Demonstrate for a Gaussian pulse the left- and right-hand sides of Equation (2.47) are equal. ∎

Exercise 2.8 Time-bandwidth product
A company is named 'time-bandwidth products'. In light of the left-hand side of Equation (2.47), what do you make of this name? ∎

2.4 The Fourier theorem

You may have heard the expression, 'You can't put a square peg into a round hole'. Well, loosely speaking, the Fourier theorem says you can make a square peg out of a round hole. For example, adding successive sines results in a square profile, Figure 2.6.

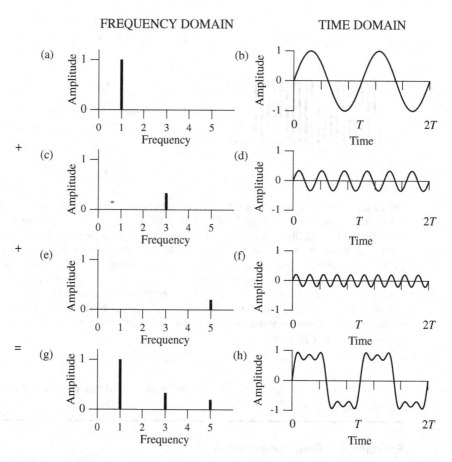

Figure 2.6 Overview of the Fourier method. Adding together a series of harmonic oscillations results in approximately a square oscillation. The concept of adding together is illustrated on the far left by the mathematical symbols of '+' and '='. On the left-hand side, the frequency domain is illustrated. The first three terms in the Fourier series are shown. The frequencies are in the ratios 1:3:5. The amplitudes – 1, 1/3 and 1/5 – are the Fourier coefficients. (a) $\sin 2\pi f_0 t$, (c) $\frac{1}{3}\sin 6\pi f_0 t$ and (e) $\frac{1}{5}\sin 10\pi f_0 t$. On the right-hand side, the time domain is given. Moving left to right on the figure is the process of Fourier synthesis. Moving right to left on the figure is the process of Fourier analysis.

The gist of the Fourier theorem is that any shape – square, triangular, whatever – can be made by combining together the circular functions of cosine and sine. To add together cosines and sines to build a profile is called **Fourier synthesis**. To take a profile and deconstruct it into its constituent cosines and sines is called **Fourier analysis**.

Let us now state the Fourier theorem a little more formally. The Fourier theorem says that any periodic phenomenon may be described as a sum of harmonic oscillations. (Even this statement is not quite precise, as there are some limitations concerning properties of the phenomenon, but they won't apply to any of the oscillations we deal with in this book.) Let me tease out a couple of points from this statement of the Fourier theorem. First, the theorem applies to periodic phenomena. Periodic phenomena occur over and over again, and by definition go on forever. So, the space across which we are working is *infinite*. Second, the number of harmonic oscillations involved may be *infinite*. Here is a second 'infinity' in stating the Fourier theorem. Of course, in practice, we may not have an infinite amount of time to study a phenomenon (the PhD needs to be written up before the money runs out). We may not have the leisure to include an infinite number of terms in our analysis. We have to modify the theorem somewhat if we are dealing with finite data sets.

The Fourier theorem is very general. The oscillation we are considering could involve any number of physical properties: force, electric field, temperature and so on. We will consider the variable in the Fourier equations to be time, t, but it could be a purely mathematical variable (perhaps the famous variable x), or a variable representing a physical quantity, such as distance. Given this generality, the Fourier theorem is widely applicable in many areas of science and engineering.

The characteristics of the cosines and sines that appear in the Fourier theorem are their frequencies and their amplitudes. (We don't need to involve the initial phase here.) The Fourier theorem connects the frequency domain with the time domain.

Let us now turn to the time domain and look carefully over one time period. Let us consider the time period, T, to run from time $t = 0$ to time $t = T$.

In building up our required function over the given period, we start with *cosines and sines that have this period*. The fundamental cosine is written as $\cos 2\pi f_0 t$, where

$$f_0 = \frac{\omega_0}{2\pi} = \frac{1}{T} \tag{2.49}$$

is the **fundamental frequency**. The second harmonic, of frequency $2f_0 = 2/T$, is also periodic over this interval, as is the third harmonic, corresponding to $3f_0$, and so on. Once T is specified, f_0 and ω_0 are known.

We are now ready to formulate the Fourier theorem mathematically. The **Fourier theorem** states any periodic function $g(t)$ may be written as the sum

$$g(t) = a_0 + \sum_{n=1}^{\infty} [a_n \cos(2\pi n f_0 t) + b_n \sin(2\pi n f_0 t)]. \tag{2.50}$$

Here, n is an index of positive integers $1, 2, 3\ldots$. In the further development I will use the more compact expressions involving ω rather than f. Then

$$g(t) = a_0 + \sum_{n=1}^{\infty}[a_n \cos(n\omega_0 t) + b_n \sin(n\omega_0 t)]. \qquad (2.51)$$

The sum may be written out explicitly as

$$g(t) = a_0 + a_1 \cos(\omega_0 t) + b_1 \sin(\omega_0 t) + a_2 \cos(2\omega_0 t) + b_2 \sin(2\omega_0 t) + \ldots \qquad (2.52)$$

The series on the right-hand side is known as the **Fourier series**.

Let us now evaluate the coefficients of the Fourier series. The evaluation of the coefficients depends on the symmetry of the cosine and sine functions and turns out to be remarkably direct.

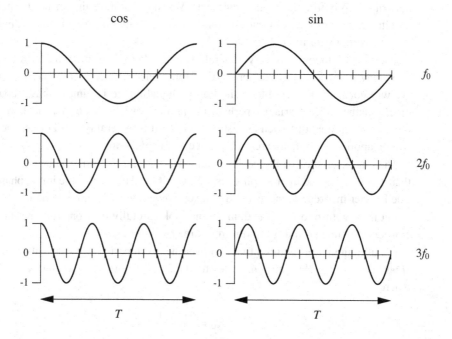

Figure 2.7 Harmonics of the cosine and sine functions over the time period T. The fundamental frequency is $f_0 = 1/T$. The frequencies depicted in successive rows are one, two and three times the fundamental frequency.

First, we will employ the average values of cosine and sine over a full period. From inspection of Figure 2.7 you should be able to see that any of the functions shown over the complete period average to zero. That is, for any value of n,

$$\int_0^T \cos(n\omega_0 t)\,dt = 0 \qquad (2.53)$$

and

$$\int_0^T \sin(n\omega_0 t)\,dt = 0. \qquad (2.54)$$

Using these results, we see that if we integrate Equation (2.51) over a full period, all the terms within the summation sign are zero, so all that remains is

$$\int_0^T g(t)dt = \int_0^T a_0 dt = a_0 T. \quad (2.55)$$

So a_0 is simply the average of $g(t)$ over the period, sometimes written as $\langle g(t) \rangle$ or as $\langle g \rangle$:

$$a_0 = \frac{1}{T} \int_0^T g(t)dt. \quad (2.56)$$

This is a constant term, not explicitly involving cosine or sine. By carrying out an integration, we reduced an equation involving an infinite series to a single term. That was fun. Let's do it again. The trick is to use the symmetry of the cosine and sine functions to produce zero integrals. Consider the product $\cos(\omega_0 t)$ and $\sin(\omega_0 t)$. With reference to Figure 2.7, without doing any calculation, you should be able to convince yourself, because the same quantities appear in different combinations of sign and cancel out, that

$$\int_0^T \cos(\omega_0 t) \sin(\omega_0 t) dt = 0 \quad (2.57)$$

and, more generally, for positive integers n and m the same or different,

$$\int_0^T \cos(n\omega_0 t) \sin(m\omega_0 t) dt = 0. \quad (2.58)$$

(If you can't see this by inspection, you are welcome to carry out the integration to prove it, using the trigonometric identities of Appendix C.) If, as here, the integral of the product of two functions is zero, we say they are **orthogonal**. Likewise, by symmetry arguments or by direct calculation, we can show the orthogonality of the cosine × cosine and sine × sine terms involving different harmonics. In other words, for $n \neq m$,

$$\int_0^T \cos(n\omega_0 t) \cos(m\omega_0 t) dt = 0, \quad (2.59)$$

$$\int_0^T \sin(n\omega_0 t) \sin(m\omega_0 t) dt = 0. \quad (2.60)$$

It seems whenever we multiply these terms together and integrate over the integral it comes to naught. What's left? Well, if $n = m$, you can verify by direct integration, or take my word for it, that

$$\int_0^T \cos(n\omega_0 t) \cos(n\omega_0 t) dt = \int_0^T \cos^2(n\omega_0 t) dt = \frac{T}{2}, \quad (2.61)$$

$$\int_0^T \sin(n\omega_0 t) \sin(n\omega_0 t) dt = \int_0^T \sin^2(n\omega_0 t) dt = \frac{T}{2}. \quad (2.62)$$

So the trick is this: by multiplying by one of the harmonics, say, $\cos(5\omega_0 t)$, and integrating over the period, all the harmonic terms $\sin(\omega_0 t)$, $\sin(2\omega_0 t)$, $\sin(3\omega_0 t)\ldots$; and

$\cos(\omega_0 t)$, $\cos(2\omega_0 t)$, $\cos(3\omega_0 t)$... *except* $\cos(5\omega_0 t)$ are knocked out and we are left with a single coefficient to evaluate:

$$\int_0^T g(t)\cos(n\omega_0 t)\,dt = \int_0^T \left(a_0 + \sum_{n=1}^{\infty}[a_n\cos(n\omega_0 t) + b_n\sin(n\omega_0 t)]\right)\cos(n\omega_0 t)\,dt$$

$$= \int_0^T a_n\cos(n\omega_0 t)\cos(n\omega_0 t)\,dt = a_n\frac{T}{2}. \tag{2.63}$$

So

$$a_n = \frac{2}{T}\int_0^T g(t)\cos(n\omega_0 t)\,dt. \tag{2.64}$$

What does this mean? Well, to get the coefficient for the $\cos(5\omega_0 t)$ term, say, we multiply the function by $\cos(5\omega_0 t)$, integrate, scale by $2/T$, and there it is. Likewise, for the sine coefficients,

$$b_n = \frac{2}{T}\int_0^T g(t)\sin(n\omega_0 t)\,dt. \tag{2.65}$$

Example 2.3 Fourier coefficient b_1

A square oscillation is defined over the period T as follows: between $t = 0$ and $t = T/2$, it takes the value 1; between $t = T/2$ and $t = T$ it takes the value -1. Determine the first sine Fourier coefficient, b_1, for this oscillation.

Solution

In solving this problem we make use of Equation (2.65) and put $n = 1$. As the function here takes two separate constant values over two separate time intervals, it is convenient to break up the integral into two parts:

$$b_1 = \frac{2}{T}\left[\int_0^{T/2} g(t)\sin(2\pi f_0 t)\,dt + \int_{T/2}^T g(t)\sin(2\pi f_0 t)\,dt\right]. \tag{2.66}$$

Now using $g(0 < t < T/2) = 1$ and $g(T/2 < t < T) = -1$, we may write

$$b_1 = \frac{2}{T}\left[\int_0^{T/2} \sin(2\pi f_0 t)\,dt - \int_{T/2}^T \sin(2\pi f_0 t)\,dt\right]. \tag{2.67}$$

The $\sin(2\pi f_0 t)$ terms each integrate up to $-\frac{1}{2\pi f_0}\cos(2\pi f_0 t)$, so

$$b_1 = \frac{2}{T}\frac{1}{2\pi f_0}\left[-\cos(2\pi f_0 t)|_0^{T/2} + \cos(2\pi f_0 t)|_{T/2}^T\right]. \tag{2.68}$$

Now we make use of Equation (2.49), $f_0 = 1/T$, to simplify matters. The leading term reduces to $1/\pi$. The value of $\cos(2\pi f_0 T/2)$ becomes $\cos(\pi) = -1$; the value of $\cos(2\pi f_0 T)$ becomes $\cos(2\pi) = 1$. So, finally,

$$b_1 = \frac{4}{\pi}. \tag{2.69}$$

Exercise 2.9 Fourier coefficient b_3

A square oscillation is defined over the period T as follows: between $t = 0$ and $t = T/2$, it takes the value 1; between $t = T/2$ and $t = T$ it takes the value -1. Determine the third sine Fourier coefficient, b_3, for this oscillation. ■

Exercise 2.10 Fourier coefficient b_5

An oscillation takes the value 1 between $t = 0$ and $t = T/2$, and the value -1 between $t = T/2$ and $t = T$. Determine the fifth sine Fourier coefficient, b_5. ■

Exercise 2.11 Fourier cosine coefficients

A square oscillation is defined over the period T as follows: between $t = 0$ and $t = T/2$, it takes the value 1; between $t = T/2$ and $t = T$ it takes the value -1. Determine the Fourier coefficients a_0, a_1 and a_2 for this oscillation. ■

2.5 Summary

2.5.1 Key terms

frequency, 4	frequency domain, 6	Fourier synthesis, 19
amplitude, 4	time domain, 6	Fourier analysis, 19
initial phase, 5	time-bandwidth theorem, 16	Fourier theorem, 19

2.5.2 Key equations

harmonic oscillation	$A = A_0 \cos(2\pi f t + \delta) = A_0 \cos\phi$	(2.18)
time-bandwidth theorem	$\Delta f \Delta t \geq 1$	(2.47)
Fourier theorem	$g(t) = a_0 + \sum_{n=1}^{\infty}[a_n \cos(2\pi n f_0 t) + b_n \sin(2\pi n f_0 t)]$	(2.50)

2.6 Table of symbols, Chapter 2

General mathematical symbols appear in Appendix B. If the unit of a quantity depends on the context, this is denoted '—'.

Symbol	Meaning	Unit
a_0	Fourier coefficient, zeroth	—
a_n	Fourier cosine coefficient, nth	—
A	general quantity	arbitrary
A_0	amplitude of oscillation	same as for A
b_n	Fourier sine coefficient, nth	—
f	frequency	Hz
f_0	frequency, fundamental	Hz
g	general function	—
m	positive integer	—
n	positive integer	—
r	modulus of complex quantity	—
R	radius	m
t	time	Hz^{-1}
t_δ	time corresponding to initial phase	Hz^{-1}
T	period	Hz^{-1}
u	real variable	—
v	real variable	—
$\tilde{w}$	complex variable	—
x	real variable; coordinate label	—
y	real variable; coordinate label	—
z'	real part of complex quantity	—
z''	imaginary part of complex quantity	—
z_1	real part of complex quantity	—
z_2	imaginary part of complex quantity	—
$\tilde{z}$	complex variable	—
$\tilde{z}*$	complex conjugate of $\tilde{z}$	—
δ	initial phase	rad
θ	general angle	rad
θ	argument of complex number	rad
ϕ	phase	rad
ω	angular frequency	rad Hz
ω_0	angular frequency, fundamental	rad Hz

3 COMBINING OSCILLATIONS

This chapter uses a lot of trigonometry. And very little else.

In the last chapter we met oscillations. In this chapter we will see what happens when we combine oscillations.

We have already seen, in passing, two examples of combining oscillations. The time-bandwidth theorem concerns oscillations with a range of frequencies and says that the spread in time goes up as the spread in frequencies goes down. The Fourier theorem says that by combining oscillations of fixed frequencies – all multiples of a fundamental frequency – an oscillation of arbitrary profile is constructed. These two examples may involve many oscillations, even an infinite number. We will take a step back in this chapter and restrict ourselves to combining two oscillations.

Underpinning all our calculations is the assumption that to combine oscillations we simply add them together. This is called the *principle of superposition*.

We begin with the simple and proceed to the complex. We will start by combining oscillations that have a lot in common, then move on to oscillations that have less in common. We saw in Chapter 2 that three key properties characterise an oscillation: *frequency*, *amplitude* and *initial phase*. We will begin by combining oscillations that have the same frequency, amplitude and initial phase, then move on to oscillations that differ only in amplitude, or only in frequency, or only in initial phase.

Two important cases stand out. If two oscillations differ only in frequency, *beats* arise. *Interference* occurs if two oscillations differ only in initial phase. Beats are important in the generation of terahertz-frequency radiation, as we will see in Chapter 7. Interference is the basis of a class of terahertz spectrometers, as we will see in Chapter 10.

So much for oscillations along a single line. Combining oscillations in different directions opens up a whole new world. We will look at combining two oscillations perpendicular to each other. *Polarisation* results. We will see that the general form of polarisation is *elliptical polarisation*, which reduces to the special cases of *linear polarisation* and *circular polarisation* in particular circumstances. We will also see that an arbitrary polarisation may be expressed using either a linear basis or a circular basis. Polarisation plays a key role in certain terahertz emitters (Chapter 7) and detectors (Chapter 9).

Learning goals

When you are done with this chapter, I expect you will be able to combine oscillations. Specifically, I expect you will be able to combine oscillations:

- that differ only in amplitude (employing the *principle of superposition*),
- that differ only in frequency (*beats*),
- that differ only in initial phase (*interference*),
- that are perpendicular to each other (*polarisation*).

3.1 Oscillations in the same direction

3.1.1 Two arbitrary oscillations

We start with two oscillations. Here they are:

$$A_1 = A_{0_1} \cos(2\pi f_1 t + \delta_1), \tag{3.1}$$

$$A_2 = A_{0_2} \cos(2\pi f_2 t + \delta_2). \tag{3.2}$$

Each oscillation is characterised by an amplitude, A_0, a frequency, f, and an initial phase, δ. In general, the amplitudes of the two oscillations may be different, the frequencies may be different, and the initial phases may be different. The subscripts '1' and '2' admit these possibilities.

To simplify things, we can write the argument of the cosine function as a single quantity, the phase, ϕ:

$$A_1 = A_{0_1} \cos\phi_1, \tag{3.3}$$

$$A_2 = A_{0_2} \cos\phi_2. \tag{3.4}$$

This is convenient if we are not interested in the details of the frequency and the time and the initial phase separately. We can unpack those from the phase as needed:

$$\phi_1 = 2\pi f_1 t + \delta_1, \tag{3.5}$$

$$\phi_2 = 2\pi f_2 t + \delta_2. \tag{3.6}$$

Instead, we can write the oscillations using the exponential notation (Section 2.2), where it is understood that at the end of the day we are only interested in the real part:

$$A_1 = A_{0_1} \exp[i(2\pi f_1 t + \delta_1)] = A_{0_1} \exp[i\phi_1] = A_{0_1} e^{i\phi_1}, \tag{3.7}$$

$$A_2 = A_{0_2} \exp[i(2\pi f_2 t + \delta_2)] = A_{0_2} \exp[i\phi_2] = A_{0_2} e^{i\phi_2}. \tag{3.8}$$

The **principle of superposition** says that we combine two oscillations by adding them. You might think this is self-evident. It is not. Not everything combines in this way. Some quantities add in quadrature: we square one, square the other, then take the square root of the result. (Two sides of a right-angle triangle combine in this way to give the hypotenuse.) In this case the dependence of the answer on the input data is not a mathematical linear relationship. In essence, the principle of superposition says the two disturbances act independently. Experimentally, this is usually the case when each disturbance is small. As the size of the disturbances grows, the principle is often overthrown. (We enter the *nonlinear* regime.) The principle of superposition is founded on observational evidence and limited in application.

An overview of the possible results of adding two oscillations is given in Figure 3.1. The three parameters A_0, f and δ are either the same or different for the two oscillations, opening up eight possibilities. In the coming sections we look at each distinct case.

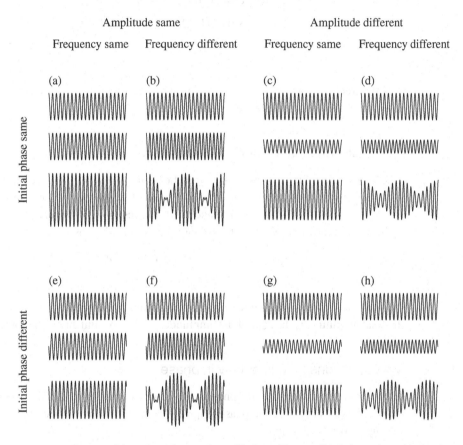

Figure 3.1 Superposition of two harmonic oscillations. The amplitude is on the vertical axis and time is on the horizontal axis. (The time axis has been drawn running from 0 to 1 second.) The three graphs in each panel display A_1, A_2 and $A_1 + A_2$. The first oscillation is $A_1 = A_{0_1} \cos(2\pi f_1 t + \delta_1)$, with always $A_{0_1} = 1$, $f_1 = 20$ Hz, and $\delta_1 = 0$. The second oscillation is $A_2 = A_{0_2} \cos(2\pi f_2 t + \delta_2)$. In panel (a) the parameters of the two oscillations are all the same. The parameters of the second oscillation differ from the first oscillation as follows: (b) $f_2 = 22$ Hz, (c) $A_{0_2} = 0.5$, (d) $A_{0_2} = 0.5$, $f_2 = 22$ Hz, (e) $\delta_2 = \pi/2$, (f) $f_2 = 22$ Hz, $\delta_2 = \pi/2$, (g) $A_{0_2} = 0.5$, $\delta_2 = \pi/2$, (h) $A_{0_2} = 0.5$, $f_2 = 22$ Hz, $\delta_2 = \pi/2$. The designations 'amplitude same' (first two columns), 'amplitude different' (second two columns), 'frequency same' (first and third columns), 'frequency different' (second and fourth columns), 'initial phase same' (first row), 'initial phase different' (second row), refer to the two initial oscillations, not necessarily to the sum.

3.1.2 Two identical oscillations

Let us apply the principle of superposition to the case when the two oscillations are the same in every respect, in other words, when the amplitudes are the same, the frequencies are the same and the initial phases are the same. This case is shown in Figure 3.1a. When a parameter is the same for both oscillations, I will drop the subscript, as it is not needed. So

$$\begin{aligned} A_1 + A_2 &= A_{0_1} \cos(2\pi f_1 t + \delta_1) + A_{0_2} \cos(2\pi f_2 t + \delta_2) \\ &= A_0 \cos(2\pi f t + \delta) + A_0 \cos(2\pi f t + \delta) \\ &= 2A_0 \cos(2\pi f t + \delta) \\ &= 2A_1 = 2A_2. \end{aligned} \qquad (3.9)$$

Thus the frequency and the initial phase of the resulting oscillation are the same as in the original oscillations while the amplitude is doubled.

3.1.3 Two oscillations that differ only in amplitude

If the frequencies of the two oscillations are the same, $f_1 = f_2 = f$, and the initial phases of the two oscillations are the same, $\delta_1 = \delta_2 = \delta$, then the two phases are the same, $\phi_1 = \phi_2 = \phi$. (Consult Equations (3.5) and (3.6) if necessary to confirm this.)

This case is shown in Figure 3.1c. Adding the two oscillations yields

$$\begin{aligned} A_1 + A_2 &= A_{0_1} \cos\phi + A_{0_2} \cos\phi \\ &= (A_{0_1} + A_{0_2}) \cos\phi. \end{aligned} \qquad (3.10)$$

The final oscillation has the same frequency and initial phase as the original oscillations. The final amplitude is the sum of the amplitudes of the constituent oscillations.

3.1.4 Two oscillations that differ only in phase

Before looking at differing initial phases (δ), we will look at differing overall phases (ϕ). The oscillations have different phases, $\phi_1 \neq \phi_2$, but the same amplitude, $A_0 = A_{0_1} = A_{0_2}$. Adding the oscillations gives

$$\begin{aligned} A_1 + A_2 &= A_0 \cos\phi_1 + A_0 \cos\phi_2 \\ &= A_0(\cos\phi_1 + \cos\phi_2). \end{aligned} \qquad (3.11)$$

We proceed using a trigonometrical formula for the sum of two cosines of arbitrary angles α and β (Appendix C),

$$\cos\alpha + \cos\beta = 2\cos\frac{\alpha-\beta}{2} \cos\frac{\alpha+\beta}{2}. \qquad (3.12)$$

Using this formula, substituting ϕ_1 for α and ϕ_2 for β, Equation (3.11) becomes:

$$A_1 + A_2 = 2A_0 \cos\frac{\phi_1-\phi_2}{2} \cos\frac{\phi_1+\phi_2}{2}. \qquad (3.13)$$

Adding two cosine functions gives a product of two cosine functions! Let's look at each cosine term in the product separately.

- $\cos\dfrac{\phi_1 - \phi_2}{2}$

    This term has as its argument half the difference of the phases. The order we write the phases doesn't matter since the cosine function is symmetric, that is, $\cos(\alpha) = \cos(-\alpha)$. We could equally as well write $\cos\dfrac{\phi_2 - \phi_1}{2}$. If the two phases happen to be the same then the term becomes $\cos 0 = 1$. If the two phases differ by an odd multiple of π (such as 3π or $-\pi$), the term becomes zero. The two oscillations completely cancel. Nothing remains.

- $\cos\dfrac{\phi_1 + \phi_2}{2}$

    This term has as its argument half the sum of the two phases. This amounts to the average (or the mean) of the two phases.

Let's introduce symbols for the sum and the difference of the two phases:

$$\Sigma \equiv \phi_1 + \phi_2, \qquad (3.14)$$

$$\Delta \equiv \phi_2 - \phi_1. \qquad (3.15)$$

Making ϕ_2 the subject of the last equation,

$$\phi_2 = \phi_1 + \Delta. \qquad (3.16)$$

Bearing in mind $\cos[(\phi_1 - \phi_2)/2] = \cos[(\phi_2 - \phi_1)/2]$, Equation (3.13) may be written:

$$A_1 + A_2 = 2A_0 \cos\frac{\Delta}{2} \cos\frac{\Sigma}{2}. \qquad (3.17)$$

How does this expression vary with Δ, the phase difference? In general, it is hard to say, since, as the phases change, Σ as well as Δ changes. However, if Σ is fixed, the expression has a simple dependence on Δ, varying as $\cos(\Delta/2)$. A particular instance of this condition is if the phases are opposite, $\phi_2 = -\phi_1$. Then $\Sigma = 0$ and $\Delta = 2\phi_2$, so $A_1 + A_2 = 2A_0 \cos(\Delta/2) = 2A_0 \cos\phi_2 = 2A_0 \cos\phi_1$.

Another simple situation is if one of the phases is fixed. Choosing ϕ_1 as the constant phase we write Equation (3.11) as

$$A_1 + A_2 = A_0[\cos\phi_1 + \cos(\phi_1 + \Delta)]. \qquad (3.18)$$

As a function of Δ, this can be read (from right to left) as a cosine function in Λ, with an initial phase given by ϕ_1, oscillating about the reference level of $\cos\phi_1$, all multiplied by A_0. This function is important in terahertz spectroscopy, as we shall see in Chapter 10. An example is plotted as Figure 3.2.

Having obtained the general expression for the sum of two oscillations that differ only in phase, let's unpack the phase in two cases (of equal amplitudes): (1) the frequencies the same but the initial phases different, Section 3.1.5 and Figure 3.1e, and (2) the initial phases the same but the frequencies different, Section 3.1.6 and Figure 3.1b.

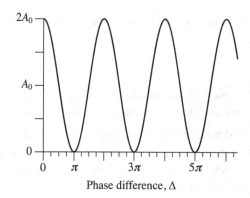

Figure 3.2 The result of combining two oscillations of equal amplitudes but different phases, $A_0 \cos \phi_1$ and $A_0 \cos \phi_2$, under the assumption that one phase is constant. The result is shown as a function of the phase difference $\Delta \equiv \phi_2 - \phi_1$. Here the constant value $\phi_1 = 0$ is used and the result is $A_0(1 + \cos \Delta)$. Note the horizontal axis is phase difference and not (as in Figure 3.1) time.

3.1.5 Two oscillations that differ only in initial phase – interference

This case is shown in Figure 3.1e. For two oscillations that differ only in initial phase, we can start from Equation (3.13), rewritten here for convenience,

$$A_1 + A_2 = 2A_0 \cos \frac{\phi_1 - \phi_2}{2} \cos \frac{\phi_1 + \phi_2}{2}. \tag{3.19}$$

Remember that ϕ is the overall phase and includes a contribution both from the frequency term and from the initial phase (Equations 3.5 and 3.6). Here the frequencies are the same, $f = f_1 = f_2$, and so the phases (Equations 3.5 and 3.6) simplify to

$$\phi_1 = 2\pi f t + \delta_1, \tag{3.20}$$

$$\phi_2 = 2\pi f t + \delta_2. \tag{3.21}$$

Inserting these expressions into Equation (3.19) gives:

$$A_1 + A_2 = 2A_0 \cos\left(\frac{\delta_1 - \delta_2}{2}\right) \cos\left(2\pi f t + \frac{\delta_1 + \delta_2}{2}\right). \tag{3.22}$$

Let's interpret this result. First we'll look at the second cosine term with the amplitude term, $A_0 \cos[2\pi f t + (\delta_1 + \delta_2)/2]$. This expression is similar to either of the constituent oscillations. It has the same amplitude and frequency. It has an initial phase that is the average of the two initial phases, $(\delta_1 + \delta_2)/2$. So this part of the result closely resembles either of the original oscillations. What's left? The term $2\cos[(\delta_1 - \delta_2)/2]$. This term does not involve the amplitude or the frequency or time but is crucial to the nature of the final oscillation. It relates to the **interference** of the two oscillations. There are two extreme types of interference.

Destructive interference

$$\delta_1 - \delta_2 = \pm n\pi, \quad n \text{ odd}, \tag{3.23}$$

In this case $\cos[(\delta_1 - \delta_2)/2] = 0$. The two oscillations neutralise each other.

Constructive interference As time goes by, the term $\cos[2\pi ft + (\delta_1 + \delta_2)/2]$ oscillates between ± 1. The term $2A_0 \cos[(\delta_1 - \delta_2)/2)]$ does not vary with time, but acts as the 'envelope' in which the time-varying part is confined. The greatest positive value it can take is $2A_0$. This is when the argument is an even multiple of π, and so the difference in initial phase is twice that:

$$\delta_1 - \delta_2 = 0, \pm 4\pi, \pm 8\pi, \ldots \qquad (3.24)$$

This is called *constructive interference*. The greatest negative value it can take is $-2A_0$. This is when the argument is an odd multiple of π, and so the difference in initial phase is twice that:

$$\delta_1 - \delta_2 = \pm 2\pi, \pm 6\pi, \pm 10\pi, \ldots \qquad \bullet \quad (3.25)$$

We have to be careful how we describe this. The outside of the envelope is now set at $-2A_0$, but since what's inside swings between ± 1, over time the range $-2A_0$ to $+2A_0$ is spanned. It is usual to refer to this case as constructive interference as well. Taking Equations (3.24) and (3.25) together, the condition for constructive interference is that the difference in initial phases is an even multiple (positive or negative) of π.

3.1.6 Two oscillations that differ only in frequency – beats

This case is shown in Figure 3.1b. When two harmonic oscillations that differ only in frequency are added together, **beats** occur. Beats are common in everyday life. For example, when two strings on a guitar are tuned almost to the same frequency and plucked together, an increase and decrease in sound level is heard. The rise and fall in volume is known as beating. Another example of acoustic beats is when two jet engines of a plane are running at slightly different speeds. The combined sound of the two engines slowly rises and falls in volume. A third example is the pattern of light and dark we see when two objects of slightly different spatial frequencies are observed at once. As I write this, I am looking at a railing on a staircase. The railing consists of vertical posts. On the far side of the staircase is a similar railing, but because it is slightly farther away, the posts appear to be slightly more closely spaced. Looked at together, sometimes the near and far posts coincide, sometimes they are close together, sometimes the near post appears right between two far posts. This pattern repeats over and over again as I run my eye along the railing. The regular bunching and spreading of the pattern constitutes spatial beats. The phenomenon of spatial beats is the basis of the vernier scale, but I imagine that topic is too old hat to interest anyone reading this.

The origin of beats is easy to explain. Think of paying two bills, one due every month from the water company, the other due every four weeks from the electricity company, starting on January 1. On January 1 you pay two bills together – not an auspicious start to the year. Your next bill, from the electricity company, is due four weeks later, on January 29. Then comes the bill from the water company, due on February 1. As time goes by, the bills keep getting further apart. So, although you pay the seventh water bill on July 1, you don't have to find the money for the eighth electricity bill until July 16 (assuming it is not a leap year). Continuing on, the bills start to arrive closer again, with bills due on August 1 and 13, September 1 and 10, October 1 and 8. The year ends with the fourteenth electricity bill due on December 31. The next day you pay the thirteenth water bill. Happy New Year.

We continue from Equation (3.13), rewritten here for convenience:

$$A_1 + A_2 = 2A_0 \cos\frac{\phi_1 - \phi_2}{2} \cos\frac{\phi_1 + \phi_2}{2}. \qquad (3.26)$$

When the initial phases of the two oscillations are the same, as we are assuming here, Equations (3.5) and (3.6) become:

$$\phi_1 = 2\pi f_1 t + \delta, \qquad (3.27)$$

$$\phi_2 = 2\pi f_2 t + \delta. \qquad (3.28)$$

Substituting these into Equation (3.26) gives:

$$A_1 + A_2 = 2A_0 \cos[2\pi\frac{f_1 - f_2}{2}t]\cos[2\pi\frac{f_1 + f_2}{2}t + \delta]. \qquad (3.29)$$

What does this mean? The part $A_0 \cos[2\pi\{(f_1 + f_2)/2\}t + \delta]$ resembles the original oscillations. It has the same amplitude (A_0) and the same initial phase (δ). The frequency ($\{f_1 + f_2\}/2$) is the average of the two initial frequencies. If the two original frequencies are similar, this expression will be very like either of the original oscillations.

What remains, $2\cos[\pi(f_1 - f_2)t]$, is also of the nature of an oscillation. Its amplitude is two, its initial phase is zero, and its frequency is $(f_1 - f_2)/2$. If the initial frequencies are close, this frequency difference will be small. This result is the basis of generating terahertz-frequency radiation at a difference frequency between two higher frequencies, as will be explained in Chapter 7.

The original two frequencies yield two characteristic frequencies in the sum: an 'inner' frequency, the average of the original frequencies, and an 'outer' frequency, related to the difference of the original frequencies. In the numerical example of Figure 3.1b, the frequencies are $f_1 = 20$ Hz, represented by 20 full cycles, and $f_2 = 22$ Hz, represented by 22 full cycles. You should be able to convince yourself by careful counting that there is the average of these, 21 full cycles, in the 'inner frequency' in the sum.

There is a subtlety regarding the 'outer frequency'. The frequency difference that appears in Equation (3.29) is $(f_1 - f_2)/2$, corresponding to $(20 - 22)/2 = -2/2 = -1$ Hz. (This negative sign is of no consequence, in view of $\cos(-\alpha) = \cos\alpha$.) In a single cycle of the envelope, $2\cos[\pi(f_1 - f_2)t]$ ranges between ± 2. Inside this is the inner component, $A_0 \cos[2\pi(\{(f_1 + f_2)/2\}t + \delta]$, which ranges between $\pm A_0$. If the outer and inner components are both negative, the same amplitude results as when they are both positive, so there are apparently two cycles of beats, rather than one, in Figure 3.1b. Because of this, the terminology *beat frequency* is used to refer to $(f_1 - f_2)$, rather than $(f_1 - f_2)/2$.

3.1 Oscillations in the same direction

Example 3.1 Acoustic beats

The fifth string on a guitar is correctly tuned to 110 Hz. What happens when it is played together with a string tuned to 116 Hz?

Solution
An intermediate or average frequency will be heard at 113 Hz. The volume will rise and fall at the difference frequency of 6 Hz. So six beats will be heard each second. (The second string can be tuned to reduce the rate of beating and bring it closer to the frequency of the first string. If this process is followed successfully then eventually the two strings will be at exactly the same frequency of 110 Hz, and no beats will be heard.)

3.1.7 Two oscillations with only frequency in common – interference

This case is shown in Figure 3.1g. It can be thought of as an extension of the case where only the amplitudes are different, Figure 3.1c, with the constraint of same initial phases removed. Or it can be thought of as an extension of the case where only the initial phases are different, Figure 3.1e, with the constraint of same amplitudes removed. This is probably a better way to look at it, as we will encounter the phenomenon of interference once more. We start by expanding each cosine term using the identity $\cos(\alpha + \beta) = \cos\alpha \cos\beta - \sin\alpha \sin\beta$ (Appendix C) and collecting like terms:

$$A_1 + A_2$$
$$= A_{0_1} \cos(2\pi f t + \delta_1) + A_{0_2} \cos(2\pi f t + \delta_2)$$
$$= A_{0_1}[\cos(2\pi f t)\cos\delta_1 - \sin(2\pi f t)\sin\delta_1] + A_{0_2}[\cos(2\pi f t)\cos\delta_2 - \sin(2\pi f t)\sin\delta_2]$$
$$= [A_{0_1}\cos\delta_1 + A_{0_2}\cos\delta_2]\cos(2\pi f t) - [A_{0_1}\sin\delta_1 + A_{0_2}\sin\delta_2]\sin(2\pi f t). \quad (3.30)$$

We can now apply in reverse $\cos(\alpha + \beta) = \cos\alpha \cos\beta - \sin\alpha \sin\beta$, provided we can find A_0' and δ' such that the terms in brackets

$$[A_{0_1}\cos\delta_1 + A_{0_2}\cos\delta_2] = A_0' \cos\delta' \quad (3.31)$$

and

$$[A_{0_1}\sin\delta_1 + A_{0_2}\sin\delta_2] = A_0' \sin\delta' \quad (3.32)$$

to yield

$$A_1 + A_2 = A_0' \cos(2\pi f t + \delta'). \quad (3.33)$$

The result is an oscillation with the same frequency as the original oscillations, but with a new amplitude and a new initial phase. The question remains – can we find values of the new amplitude and phase? The answer is yes. First, to solve for A_0', we eliminate δ'

by squaring Equation (3.31) and squaring Equation (3.32) and adding them together:

$$(A'_0 \cos \delta')^2 + (A'_0 \sin \delta')^2 = A'^2_0 (\cos^2 \delta' + \sin^2 \delta') = A'^2_0$$
$$= [A_{0_1} \cos \delta_1 + A_{0_2} \cos \delta_2]^2 + [A_{0_1} \sin \delta_1 + A_{0_2} \sin \delta_2]^2$$
$$= A^2_{0_1} + A^2_{0_2} + 2A_{0_1} A_{0_2} \cos(\delta_1 - \delta_2). \tag{3.34}$$

(In establishing this result, I have used the identities $\cos^2 \alpha + \sin^2 \alpha = 1$ and $\cos(\alpha - \beta) = \cos \alpha \cos \beta + \sin \alpha \sin \beta$, Appendix C.) So

$$A'_0 = \sqrt{A^2_{0_1} + A^2_{0_2} + 2A_{0_1} A_{0_2} \cos(\delta_1 - \delta_2)}. \tag{3.35}$$

The amplitudes add in quadrature, with the addition of the 'interference term' amounting to $2A_{0_1} A_{0_2} \cos(\delta_1 - \delta_2)$; if the initial phases are equal, the interference term vanishes (corresponding to constructive interference). If the amplitudes are equal, we may utilise $\cos^2 \alpha = (\cos 2\alpha + 1)/2$ (Appendix C) to write $A'_0 = A_0 \sqrt{2 + 2\cos(\delta_1 - \delta_2)} = 2A_0 \cos[(\delta_1 - \delta_2)/2]$, as appears in Equation (3.22).

What about δ'? We eliminate A'_0 by dividing Equation (3.32) by Equation (3.31):

$$\frac{A_{0_1} \sin \delta_1 + A_{0_2} \sin \delta_2}{A_{0_1} \cos \delta_1 + A_{0_2} \cos \delta_2} = \frac{A'_0 \sin \delta'}{A'_0 \cos \delta'} = \tan \delta'. \tag{3.36}$$

Hence

$$\delta' = \arctan \left[\frac{A_{0_1} \sin \delta_1 + A_{0_2} \sin \delta_2}{A_{0_1} \cos \delta_1 + A_{0_2} \cos \delta_2} \right]. \tag{3.37}$$

3.1.8 Two oscillations with only initial phase in common – beat-like

This case is shown in Figure 3.1d. It can be thought of as an extension of the case where only the amplitudes are different, Figure 3.1c, with the constraint of same frequencies removed. Or it can be thought of as an extension of the case where only the frequencies are different, Figure 3.1b, with the constraint of same amplitudes removed. This may be a better way to look at it, as we will encounter the phenomenon of beats once more.

We proceed as in Section 3.1.7. I won't here present in detail each step as I did there. We expand each cosine and collect like terms:

$$A_{0_1} \cos(2\pi f_1 t + \delta) + A_{0_2} \cos(2\pi f_2 t + \delta)$$
$$= [A_{0_1} \cos(2\pi f_1 t) + A_{0_2} \cos(2\pi f_2 t)] \cos \delta - [A_{0_1} \sin(2\pi f_1 t) + A_{0_2} \sin(2\pi f_2 t)] \sin \delta. \tag{3.38}$$

We seek A''_0 and f'' such that

$$[A_{0_1} \cos(2\pi f_1 t) + A_{0_2} \cos(2\pi f_2 t)] = A''_0 \cos 2\pi f'' t \tag{3.39}$$

and

$$[A_{0_1} \sin(2\pi f_1 t) + A_{0_2} \sin(2\pi f_2 t)] = A''_0 \sin 2\pi f'' t, \tag{3.40}$$

to yield

$$A_1 + A_2 = A''_0 \cos(2\pi f'' t + \delta). \tag{3.41}$$

Solving Equations (3.39) and (3.40) for A_0'' and f'' gives

$$A_0'' = \sqrt{A_{0_1}^2 + A_{0_2}^2 + 2A_{0_1}A_{0_2}\cos[2\pi(f_1 - f_2)t]}. \tag{3.42}$$

$$2\pi f'' t = \arctan\left[\frac{A_{0_1}\sin(2\pi f_1 t) + A_{0_2}\sin(2\pi f_2 t)}{A_{0_1}\cos(2\pi f_1 t) + A_{0_2}\cos(2\pi f_2 t)}\right]. \tag{3.43}$$

This is messy. The key point to see is that the expression for the amplitude involves an average amplitude term, $A_{0_1}^2 + A_{0_2}^2$, that does not completely cancel with the interference term, $2A_{0_1}A_{0_2}\cos[2\pi(f_1 - f_2)t]$. The overall combination is therefore beat-like, but the modulation does not reduce the amplitude all the way to zero; see Figure 3.1d.

3.1.9 Two oscillations with only amplitude in common – more beats

This case is shown in Figure 3.1f. In Equation (3.13), rewritten here for convenience,

$$A_1 + A_2 = 2A_0 \cos\frac{\phi_1 - \phi_2}{2}\cos\frac{\phi_1 + \phi_2}{2}, \tag{3.44}$$

we put

$$\phi_1 = 2\pi f_1 t + \delta_1, \tag{3.45}$$

$$\phi_2 = 2\pi f_2 t + \delta_2, \tag{3.46}$$

to yield:

$$A_1 + A_2 = 2A_0 \cos[2\pi\frac{f_1 - f_2}{2}t + \frac{\delta_1 - \delta_2}{2}]\cos[2\pi\frac{f_1 + f_2}{2}t + \frac{\delta_1 + \delta_2}{2}]. \tag{3.47}$$

The result is rather similar to that for beats as we first developed them, Equation (3.29). The only difference is a different initial phase in the inner frequency term, $(\delta_1 + \delta_2)/2$, and a new initial phase, $(\delta_1 - \delta_2)/2$, in the outer frequency term. Thus both the inner and outer oscillations are offset in phase in Figure 3.1f relative to Figure 3.1b.

3.1.10 Two oscillations with nothing in common

This case is shown in Figure 3.1h. In general, it can involve a mixture, complete or incomplete, of the phenomena we have encountered: interference and beats.

Exercise 3.1 Adding oscillations

Give the expressions for the combination of two oscillations in the eight cases shown in Figure 3.1 for the parameters: $A_{0_1} = 1$, $f_1 = 1$ THz, $\delta_1 = 0$, $A_{0_2} = 0.5$, $f_2 = 1.1$ THz, $\delta_2 = \pi/2$. ∎

Exercise 3.2 Frequency domain

Prepare eight sketches, similar to Figure 3.1a–h, but where the two original oscillations and their sum are represented in the frequency domain. ∎

3.2 Oscillations in two dimensions

Now we will step it up a notch. Instead of dealing with oscillations in only one dimension, we will turn to oscillations in two dimensions. We will start with the simplest case, two oscillations of the same amplitude. Then we will look at the more general case, two oscillations of different amplitudes. There are a couple of ways to do this. We will also use a compact notation to describe two-dimensional oscillations. *Throughout this section, we assume that the two oscillations have the same frequency.*

3.2.1 Two perpendicular oscillations: same amplitude

Up until now we have considered oscillations in one dimension only. We labelled that dimension x. Now we will introduce a second dimension. We will label this y. We will assume that y is measured in a direction perpendicular to x. We can specify any position in the two-dimensional x-y plane by giving values for the two coordinates x and y.

We will now combine two oscillations, one in the x direction, one in the y direction. We are assuming the frequency f (and so $\omega = 2\pi f$) is the same for each direction and the amplitude A_0 is the same for each direction. To start as simply as possible, let's now assume the oscillations have the same initial phase, δ. Then

$$x = A_0 \cos(2\pi f t + \delta), \tag{3.48}$$

$$y = A_0 \cos(2\pi f t + \delta). \tag{3.49}$$

These are the same equation! So $y = x$. This is the equation of a straight line, of slope 1, running through the origin. Adding two harmonic oscillations perpendicular to each other has resulted in a linear harmonic oscillation. This is illustrated in Figure 3.3a. We

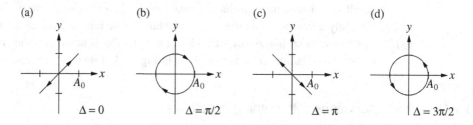

Figure 3.3 Combining two perpendicular oscillations of the same frequency and the same amplitude (A_0). The nature of the resulting oscillation depends on the difference in initial phase of the two original oscillations, $\Delta = \delta_y - \delta_x$. (a) $\Delta = 0$. The oscillation is along the line $y = x$ between the limits $\pm A_0$. (b) $\Delta = \pi/2$. The oscillation is clockwise around a circle of radius A_0. (c) $\Delta = \pi$. The oscillation is along the line $y = -x$ between the limits $\pm A_0$. (d) $\Delta = 3\pi/2$. The oscillation is anticlockwise around a circle of radius A_0.

also end up with a straight line if the second oscillation is exactly out of phase with the first, that is, if $y = A_0 \cos(2\pi f t + \delta + \pi)$. Then $y = -x$, corresponding to a straight line of slope -1 and running through the origin. This is perpendicular to the previous line, $y = x$, and is shown in Figure 3.3c.

3.2 Oscillations in two dimensions

The next simplest case is when the two perpendicular oscillations of the same frequency and amplitude differ in phase by a quarter of a cycle, or by $\pi/2$ radians:

$$x = A_0 \cos(2\pi ft + \delta), \tag{3.50}$$

$$y = A_0 \cos(2\pi ft + \delta + \pi/2) = -A_0 \sin(2\pi ft + \delta). \tag{3.51}$$

(In the expression for y, I have used the identity $\cos(\alpha + \pi/2) = -\sin \alpha$, Appendix C.) Equations (3.50) and (3.51) are directly related to motion on a circle, as we have discussed previously (Section 2.1). We eliminate time from the two equations via the identity $\sin^2 \alpha + \cos^2 \alpha = 1$ (Appendix C) with $\alpha = 2\pi ft + \delta$. Hence

$$x^2 + y^2 = A_0^2. \tag{3.52}$$

This is the equation for a circle of radius A_0, centred at the origin, Figure 3.3b. The circle is traced out in the clockwise direction. If the phase difference is reversed, that is, if $y = A_0 \cos(2\pi ft + \delta - \pi/2) = A_0 \sin(2\pi ft + \delta)$, we obtain the same equation for the path, $x^2 + y^2 = A_0^2$, Figure 3.3d, but the circle is now traced in the anticlockwise direction. The same result holds for $y = A_0 \cos(2\pi ft + \delta + 3\pi/2)$.

Let's consider the most general case of combining two perpendicular oscillations of the same frequency and amplitude. We assume the two initial phases are different and denote them δ_x and δ_y. We use the relation $\cos(\alpha + \beta) = \cos \alpha \cos \beta - \sin \alpha \sin \beta$ (Appendix C) to write

$$x = A_0 \cos(2\pi ft + \delta_x) = A_0[\cos(2\pi ft)\cos(\delta_x) - \sin(2\pi ft)\sin(\delta_x)], \tag{3.53}$$

$$y = A_0 \cos(2\pi ft + \delta_y) = A_0[\cos(2\pi ft)\cos(\delta_y) - \sin(2\pi ft)\sin(\delta_y)]. \tag{3.54}$$

We can employ the same strategy as a moment ago to eliminate t from these equations, that is, use the identity $\sin^2 \alpha + \cos^2 \alpha = 1$, here with $\alpha = 2\pi ft$. A few lines of working are involved to obtain:

$$x^2 - 2xy \cos(\delta_y - \delta_x) + y^2 = A_0^2 \sin^2(\delta_y - \delta_x). \tag{3.55}$$

The initial phases δ_x and δ_y do not appear in this equation separately, only their difference, $\delta_y - \delta_x$. I'll use the symbol Δ to stand for the difference in initial phase:

$$\Delta \equiv \delta_y - \delta_x. \tag{3.56}$$

Hence Equation (3.55) may be written as

$$x^2 - 2xy \cos \Delta + y^2 = A_0^2 \sin^2 \Delta. \tag{3.57}$$

Equation (3.57) is the equation of an ellipse, although you may not recognise it as such. We may *define* an ellipse in this very way: *an ellipse is the path obtained by combining two perpendicular harmonic motions of the same frequency and arbitrary initial phases.* From this viewpoint, the straight line ($\Delta = 0$ or π) and the circle($\Delta = \pi/2$ or $-\pi/2$) may be seen to be special cases of the general case of the ellipse. We refer to the resulting shape – line, circle or ellipse – as the **polarisation** of the oscillation.

Although Equation (3.57) represents an ellipse, the axes of the ellipse do not run along the x and y axes. (In fact, the ellipse axes make angles of $\pm\pi/4$ with respect to the x axis.) We can simplify the appearance of Equation (3.57) by getting rid of the cross term involving xy. We can do this by transforming the axes x and y to new axes X and Y defined by

$$x = \frac{1}{\sqrt{2}}[X - Y], \tag{3.58}$$

$$y = \frac{1}{\sqrt{2}}[X + Y]. \tag{3.59}$$

(X and Y correspond to x and y rotated through $\pi/4$ in the anticlockwise sense.) After a few lines of working, we may rewrite Equation (3.57) in the more usual form of the ellipse,

$$\left(\frac{X}{a}\right)^2 + \left(\frac{Y}{b}\right)^2 = 1. \tag{3.60}$$

Here

$$a = A_0 \sqrt{1 + \cos \Delta} \tag{3.61}$$

and

$$b = A_0 \sqrt{1 - \cos \Delta} \tag{3.62}$$

are each half the length of one axis of the ellipse. The ellipse is shown in Figure 3.4.

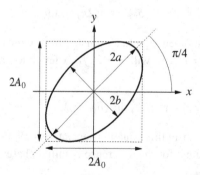

Figure 3.4 Combining two perpendicular oscillations of the same frequency and the same amplitude (A_0) and initial phase difference $\Delta = \delta_y - \delta_x = \pi/3$. One axis of the ellipse makes an angle of $\pi/4$ with the x axis, as shown, regardless of the value of Δ. The lengths of half the ellipse axes are $a = A_0 \sqrt{1 + \cos \Delta}$ and $b = A_0 \sqrt{1 - \cos \Delta}$.

Exercise 3.3 Obtaining the ellipse equation

(*This exercise assumes some knowledge of trigonometry.*) Derive Equation (3.55) from Equations (3.53) and (3.54).

Hint: Multiply Equation (3.53) by $\cos \delta_y$ and Equation (3.54) by $\cos \delta_x$, then subtract the results to eliminate $\cos(2\pi ft)$ and obtain an expression for $\sin(2\pi ft)$. A similar ploy yields an expression for $\cos(2\pi ft)$. Then eliminate t via $\sin^2(2\pi ft) + \cos^2(2\pi ft) = 1$.

∎

Exercise 3.4 Rotating the ellipse
(*This exercise assumes good knowledge of trigonometry.*) Derive Equation (3.60) from Equations (3.57), (3.58) and (3.59). Confirm the expressions for a and b given in Equations (3.61) and (3.62). ∎

3.2.2 Two perpendicular oscillations: different amplitudes

Let's now turn to a more general case still: two perpendicular oscillations with the same frequency but different amplitudes and arbitrary phases. I will denote the two amplitudes A_x and A_y. (I could use x_0 and y_0 instead, but I think using A_x and A_y is clearer here.)

$$x = A_x \cos(2\pi ft + \delta_x) = A_x[\cos(2\pi ft)\cos(\delta_x) - \sin(2\pi ft)\sin(\delta_x)], \quad (3.63)$$

$$y = A_y \cos(2\pi ft + \delta_y) = A_y[\cos(2\pi ft)\cos(\delta_y) - \sin(2\pi ft)\sin(\delta_y)]. \quad (3.64)$$

Employing the same strategy as before to eliminate t,

$$\left(\frac{x}{A_x}\right)^2 - 2\left(\frac{x}{A_x}\right)\left(\frac{y}{A_y}\right)\cos\Delta + \left(\frac{y}{A_y}\right)^2 = \sin^2\Delta. \quad (3.65)$$

As before, the two individual initial phases δ_x and δ_y are not needed in the final equation, but only their difference, $\Delta = \delta_y - \delta_x$. After an appropriate coordinate transformation, this equation may be written in the usual form of the ellipse, $(X/a)^2 + (Y/b)^2 = 1$. The angle, α, through which the x-y axes need to be rotated (in the anticlockwise direction) to align with the X-Y axes is determined from

$$\tan 2\alpha = \frac{2A_x A_y}{A_x^2 - A_y^2} \cos\Delta. \quad (3.66)$$

Take care: the equation refers to 2α, but the angle through which the axes need to be rotated is half this, α. Also note that the calculation of the axis rotation angle involves not only the relative size of the amplitudes, but also the relative initial phase. This is different from the case of identical amplitudes; then the ellipse axes were always along $y = \pm x$, regardless of the initial phases.

We may determine the length of the semimajor axis, a, and the semiminor axis, b, of the rotated ellipse. The result depends on the amplitudes of the two oscillations, A_x and A_y, as well as the relative phase, Δ, and is a little involved:

$$a^2 = \frac{A_x^2 + A_y^2}{2} + \left[\left(\frac{A_x^2 - A_y^2}{2}\right)^2 + (A_x A_y \cos\Delta)^2\right]^{1/2}, \quad (3.67)$$

$$b^2 = \frac{A_x^2 + A_y^2}{2} - \left[\left(\frac{A_x^2 - A_y^2}{2}\right)^2 + (A_x A_y \cos\Delta)^2\right]^{1/2}. \quad (3.68)$$

An example of adding two perpendicular oscillations of different phases and amplitudes is given in Figure 3.5.

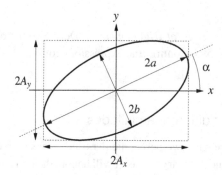

Figure 3.5 Combining two perpendicular oscillations of the same frequency and the different amplitudes ($A_x = 1.5 \times A_y$) and initial phase difference $\Delta = \delta_y - \delta_x = \pi/3$.

Exercise 3.5 Obtaining a more general ellipse equation

(This exercise assumes some knowledge of trigonometry.) Derive Equation (3.65) from Equations (3.63) and (3.64).

Hint: Divide Equation (3.63) by A_x and Equation (3.64) by A_y. The right-hand sides of these equations are the same as the right-hand sides of Equations (3.53) and (3.54) divided by A_0. Now proceed as in Exercise 3.3. Alternatively, proceed directly by replacing x with x/A_x, y with y/A_y and A_0 with 1 in Equation (3.57). ∎

Exercise 3.6 Rotating the general ellipse

(This exercise assumes good knowledge of trigonometry.) Derive Equation (3.66).

Hint: In Equation (3.65) make the substitutions

$$x = X \cos \alpha - Y \sin \alpha, \tag{3.69}$$

$$y = X \sin \alpha + Y \cos \alpha. \tag{3.70}$$

Then setting the cross term (in XY) to zero requires

$$(A_x^2 - A_y^2) \sin 2\alpha = 2 A_x A_y \cos \Delta \cos 2\alpha, \tag{3.71}$$

directly leading to the result. ∎

Exercise 3.7 Semimajor and semiminor axes

(Easy.) Show that for the case of equal amplitudes, $A_x = A_y = A_0$, Equations (3.67) and (3.68) reduce to Equations (3.61) and (3.62). ∎

Exercise 3.8 Obtaining the semimajor and semiminor axes of the general ellipse

(Challenging.) Derive Equations (3.67) and (3.68). ∎

Exercise 3.9 Inverting the general ellipse equations

(Challenging.) Show that a general ellipse centred at the origin, rotated at a specified angle from the x axis, and of specified semimajor and semiminor axis lengths, can be obtained by combining harmonic oscillations of the same frequency in the x and y directions.

In other words, invert Equations (3.66), (3.67) and (3.68) to obtain expressions for A_x, A_y and Δ, given α, a and b. ∎

3.2.3 Two counter-rotating circular oscillations

Let's look at oscillations in two-dimensional space from another perspective. Depending on how your brain functions, how your thoughts follow one from the other, or how the cogs in your mind turn, you may find this perspective more or less helpful than the approach of Section 3.2.1.

In this new perspective, we will not start with oscillations along a single dimension, but motion in a circle. We have met motion in a circle before (Section 2.1). We saw that a harmonic oscillation could be thought of as the projection of uniform circular motion onto a particular axis. Now we meet something connected, but different. We will see that a harmonic oscillation may be regarded as the combination of two counter-rotating circular motions. Adding two circles will result in a straight line.

Let's start with motion around a circle with initial phase zero. From Equations (2.5) and (2.6), the x and y coordinates are

$$x_{\circlearrowleft} = A_0 \cos 2\pi ft, \tag{3.72}$$

$$y_{\circlearrowleft} = A_0 \sin 2\pi ft. \tag{3.73}$$

The direction of motion is crucial to this discussion. Let's check the direction of the rotation. At $t = 0$, the coordinates are $(\cos 0, \sin 0) = (1, 0)$. Shortly after, when $2\pi ft$ has a small, positive value, x will have decreased slightly and y will have increased slightly. This initial movement, to the left and up in the x-y plane, corresponds to anticlockwise motion. This is why I have explicitly labelled the coordinates with the subscript $\circlearrowleft$, standing for anticlockwise.

How do we represent a clockwise rotation? There are two ways, you can choose either. You might take time to run backwards. Or you might think of the frequency as being negative. Either way, the term $2\pi ft$ changes sign:

$$x_{\circlearrowright} = A_0 \cos(-2\pi ft), \tag{3.74}$$

$$y_{\circlearrowright} = A_0 \sin(-2\pi ft). \tag{3.75}$$

Let's see what happens when we add the anticlockwise and clockwise motions. Making use of $\cos(-\alpha) = \cos \alpha$ and $\sin(-\alpha) = -\sin \alpha$ (Appendix C),

$$x_{\circlearrowleft} + x_{\circlearrowright} = A_0(\cos(2\pi ft) + \cos(-2\pi ft)) = 2A_0 \cos(2\pi ft), \tag{3.76}$$

$$y_{\circlearrowleft} + y_{\circlearrowright} = A_0(\sin(2\pi ft) + \sin(-2\pi ft)) = 0. \tag{3.77}$$

So adding a clockwise rotation to an anticlockwise rotation results in harmonic motion along the x axis! There is an exclamation mark ending the last sentence, as this is a remarkable result. The y components cancel completely (Equation 3.77), only leaving the x components, which add to give a harmonic oscillation (Equation 3.76). This is illustrated in Figure 3.6. This is different from, and perhaps more satisfying than, the burden of Section 2.1, that the projection of circular motion onto a line gives a harmonic oscillation. We can make an oscillation restricted to the y axis by differencing the opposite

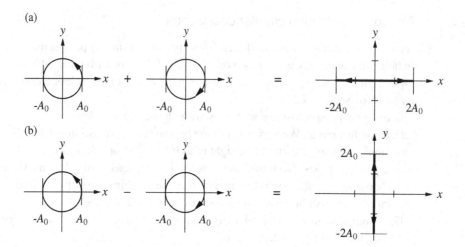

Figure 3.6 (a) Adding two counter-rotating circular oscillations results in a linear oscillation. (b) Differencing the same two counter-rotating circular oscillations results in a linear oscillation perpendicular to the first.

rotations:

$$x_{\circlearrowleft} - x_{\circlearrowright} = A_0(\cos(2\pi ft) - \cos(-2\pi ft)) = 0, \qquad (3.78)$$

$$y_{\circlearrowleft} - y_{\circlearrowright} = A_0(\sin(2\pi ft) - \sin(-2\pi ft)) = 2A_0 \sin(2\pi ft). \qquad (3.79)$$

We have come full circle (pun intended). In Section 3.2.1 we saw that two one-dimensional oscillations along two perpendicular axes could combine to produce motion in a circle. Now we have seen that two two-dimensional circular motions can combine to give motion along a single axis. So, starting with two circular motions, we can construct perpendicular straight-line motions, which we can then use to produce arbitrary ellipses, as in Section 3.2.2, to our hearts' content. The general ellipse (Figure 3.5) can be realised starting either with straight-line or circular oscillations. One approach is no more fundamental than the other.

3.2.4 Compact notation for two-dimensional oscillations and rotation

In combining oscillations in two dimensions, I have been writing out the x and y harmonic oscillations separately, for example, in Equations (3.63) and (3.64) as

$$x = A_x \cos(2\pi ft + \delta_x), \qquad (3.80)$$

$$y = A_y \cos(2\pi ft + \delta_y). \qquad (3.81)$$

A quantity that has two (or more) components and combines as displacements do (by the parallelogram rule) is called a **vector**. We can write the two components slightly more compactly in a column vector, where it is understood that the top entry is the x

3.2 Oscillations in two dimensions

value and the bottom entry is the y value:

$$\begin{bmatrix} x \\ y \end{bmatrix} = \begin{bmatrix} A_x \cos(2\pi ft + \delta_x) \\ A_y \cos(2\pi ft + \delta_y) \end{bmatrix}. \tag{3.82}$$

Alternatively, we could write the components in a horizontal format as a row vector:

$$(x, y) = (A_x \cos(2\pi ft + \delta_x), A_y \cos(2\pi ft + \delta_y)). \tag{3.83}$$

Another notation still uses the unit vectors in the x and y directions, $\hat{\mathbf{x}}$ and $\hat{\mathbf{y}}$. The bold font indicates a vector. The hat (or caret) represents a unit length. Thus

$$(x, y) = A_x \cos(2\pi ft + \delta_x)\hat{\mathbf{x}} + A_y \cos(2\pi ft + \delta_y)\hat{\mathbf{y}}. \tag{3.84}$$

The notation I want to develop in more detail is based on the concept of the complex number $\tilde{z}$, defined in Section 2.2. Let's first apply this to the case of equal amplitudes, $A_x = A_y$. Putting Equations (3.80) and (3.81) into Equation (2.23),

$$\tilde{z} = A_x \cos(2\pi ft + \delta_x) + iA_y \cos(2\pi ft + \delta_y). \tag{3.85}$$

Compare this with Equation (2.44),

$$\tilde{z} = A_0 e^{i\phi} = A_0 \cos\phi + iA_0 \sin\phi, \tag{3.86}$$

where $\phi = 2\pi ft + \delta$. Now, here's the thing: when previously writing $A = A_0 e^{i\phi}$ (Equation 2.46), I said we understood the oscillation to be encoded in the real part, $A = \text{Re}(A_0 e^{i\phi}) = A_0 \cos\phi$ (Equation 2.44) and the rest, $\text{Im}(A_0 e^{i\phi}) = A_0 \sin\phi$, was superfluous. That was so in describing oscillations along a line. Now, in writing $\tilde{z} = A_0 e^{i\phi}$, the term $A_0 \sin\phi$ turns out to be useful as the perpendicular component in the representation of circular motion.

Introducing rotation

Let's introduce a second complex number, associated with another angle, β:

$$e^{i\beta} = \cos\beta + i\sin\beta. \tag{3.87}$$

This is of unit magnitude (as you can see by adding the components in quadrature, $\cos^2\beta + \sin^2\beta = 1$). Let's take the idea further. In doing so, we will see how the exponential notation comes into its own. It follows from the way we multiply using indices, $a^m \times a^n = a^{n+m}$ (Appendix C). Let's see what happens when we multiply our first complex number by the second:

$$\tilde{z} e^{i\beta} = A_0 e^{i\phi} e^{i\beta} = A_0 e^{i(\phi+\beta)}. \tag{3.88}$$

The result is a quantity of the same size as $\tilde{z}$, namely A_0, but at a different angle to the x axis, namely $\phi + \beta$. The effect of multiplying by $e^{i\beta}$ is to rotate the original quantity through an angle of β. We will be using this remarkable property over and over again. It is illustrated in Figure 3.7.

Let us take a more pedestrian look at this rotation. The angle β is zero on the x axis. So we may write the unit vector along the x axis as

$$\hat{\mathbf{x}} = e^{i0} = \cos 0 + i\sin 0 = 1 + i0 = 1. \tag{3.89}$$

44 COMBINING OSCILLATIONS

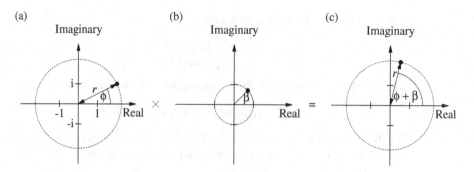

Figure 3.7 Multiplication of two complex numbers: $\tilde{z} \times e^{i\beta} = re^{i\phi} \times e^{i\beta} = re^{i(\phi+\beta)}$. Here $\tilde{z} = 2 + i$, so $r = \sqrt{5}$ and $\phi = \arctan(1/2)$; $\beta = \pi/4$. The overall effect is to rotate $\tilde{z}$ anticlockwise about the origin through the angle β. This holds for any value of β.

The angle β is $\pi/2$ on the y axis. So we may write the unit vector along the y axis as

$$\hat{y} = e^{i\pi/2} = \cos \pi/2 + i \sin \pi/2 = 0 + i1 = i. \tag{3.90}$$

In these two equations we link the geometric x-y plane and its unit vectors $\hat{x}$ and $\hat{y}$ with the complex plane and its real and imaginary axes. We will move between these seamlessly from now on.

Let's apply the quantity $e^{i\beta}$ to each unit vector, and call the results $\hat{X}$ and $\hat{Y}$:

$$\hat{X} = e^{i\beta}\hat{x} = (\cos\beta + i\sin\beta)\hat{x} = \cos\beta\hat{x} + \sin\beta\hat{y}. \tag{3.91}$$

$$\hat{Y} = e^{i\beta}\hat{y} = i\cos\beta + i(i\sin\beta) = i\cos\beta - \sin\beta = -\sin\beta\hat{x} + \cos\beta\hat{y}. \tag{3.92}$$

You may recognise these terms as matrix elements for rotation by an angle of β anti-clockwise in the x-y plane.

$$\begin{bmatrix} \hat{X} \\ \hat{Y} \end{bmatrix} = \begin{bmatrix} \cos\beta & \sin\beta \\ -\sin\beta & \cos\beta \end{bmatrix} \begin{bmatrix} \hat{x} \\ \hat{y} \end{bmatrix}. \tag{3.93}$$

The inverse matrix lets us express $\hat{x}$ and $\hat{y}$ in terms of $\hat{X}$ and $\hat{Y}$:

$$\begin{bmatrix} \hat{x} \\ \hat{y} \end{bmatrix} = \begin{bmatrix} \cos\beta & -\sin\beta \\ \sin\beta & \cos\beta \end{bmatrix} \begin{bmatrix} \hat{X} \\ \hat{Y} \end{bmatrix}. \tag{3.94}$$

We have already used this result for rotation through $\pi/4$ (Equations 3.58 and 3.59) and for rotation through α (Equations 3.69 and 3.70).

3.2.5 Counter-rotating circular oscillations, compact notation

Our appetites have already been whetted with what we can do in adding circular oscillations in Section 3.2.3. Now let's really tuck into it, using the compact notation just introduced in Section 3.2.4. In particular, we will make use of the fact that multiplication by $e^{i\beta}$ corresponds to rotation through an angle of β.

3.2 Oscillations in two dimensions

At the end, we will come to the combination of two opposite rotations at the same frequency but of arbitrary amplitudes and initial phases. I introduce the nomenclature 'left' to refer to an anticlockwise rotation and 'right' to refer to a clockwise rotation, and the subscripts 'L' and 'R' to refer to these. (These play an identical role in a new context to the subscripts ↺ and ↻ in Section 3.2.3.)

$$A = A_L e^{i(2\pi ft + \delta_L)} + A_R e^{-i(2\pi ft + \delta_R)}. \tag{3.95}$$

First off, let's consider only a left amplitude A_L and zero right amplitude ($A_R = 0$). This corresponds to a circle traced out in the left-hand, or anticlockwise, direction:

$$A = A_L e^{i(2\pi ft + \delta_L)}. \tag{3.96}$$

At time $t = 0$ the position is

$$A_L e^{i\delta_L} = A_L(\cos\delta_L + i\sin\delta_L). \tag{3.97}$$

This point is on the line $y/x = \sin\delta_L / \cos\delta_L = \tan\delta_L$, and a distance A_L from the origin. To start the motion on the x axis requires $\delta_L = 0$; motion starting on the y axis corresponds to $\delta_L = \pi/2$. Likewise, for clockwise, or right-hand, rotation,

$$A = A_R e^{-i(2\pi ft + \delta_R)}. \tag{3.98}$$

The $t = 0$ position is

$$A_R e^{-i\delta_R} = A_R(\cos\delta_R - i\sin\delta_R). \tag{3.99}$$

This is on the line $y/x = -\sin\delta_R / \cos\delta_R = -\tan\delta_R$, and a distance A_R from the origin. Notice that, because of the negative sign in the exponent, the same values of initial phase $\delta_L = \delta_R$ do not correspond to the same starting point (an exception is if each initial phase is a multiple of π).

Thus, by varying the values of A_L, A_R, δ_L and δ_R, we may obtain a circular path centred at the origin, of any radius we wish, with the motion starting anywhere on the circle we wish, and proceeding in either direction.

Initial phases zero

Let's now add together these circular functions to obtain a straight line. We did this before (Section 3.2.3), let's do it again, using our new notation, and paying close attention to the initial phase. Let's set each amplitude to A_0 and each initial phase to zero:

$$A_L e^{i(2\pi ft + \delta_L)} + A_R e^{-i(2\pi ft + \delta_R)} = A_0 e^{i(2\pi ft)} + A_0 e^{-i(2\pi ft)} = 2A_0 \cos(2\pi ft). \tag{3.100}$$

This is motion along the x axis, with amplitude twice that of the constituent rotations. The complex notation has collapsed to a real quantity. I won't pause here to again marvel that two circles make a straight line (Figure 3.6).

Initial phases equal

Now let's set each phase to the same value, δ.

$$A_L e^{i(2\pi ft + \delta_L)} + A_R e^{-i(2\pi ft + \delta_R)} = A_0 e^{i(2\pi ft + \delta)} + A_0 e^{-i(2\pi ft + \delta)} = 2A_0 \cos(2\pi ft + \delta). \tag{3.101}$$

The motion is still restricted to the x axis, with an amplitude twice that of the constituent rotations, but starts at a different position. By varying the common value of δ between 0 and 2π, we may start the motion anywhere along the x axis and heading in either direction.

Initial phases opposite

Now let's set the phases to opposite values, $\delta = \delta_L = -\delta_R$. This again results in a straight line, but no longer fixed to the x axis:

$$\begin{aligned} A_0 e^{i(2\pi ft+\delta)} + A_0 e^{-i(2\pi ft-\delta)} &= A_0 e^{i\delta}\left(e^{i(2\pi ft)} + e^{-i(2\pi ft)}\right) \\ &= 2A_0 e^{i\delta} \cos(2\pi ft) \\ &= (\cos\delta + i\sin\delta) 2A_0 \cos(2\pi ft). \end{aligned} \quad (3.102)$$

We have managed to separate out the straight-line motion part ($2A_0 \cos(2\pi ft)$) from the angular part ($e^{i\delta}$). The motion is along a straight line making a tangent of $y/x = \sin\delta/\cos\delta = \tan\delta$ with the x axis.

Initial phases recast

We have gained some insight into the role of the initial phase when adding two circular oscillations. The part that is in common controls where on the overall cycle the motion begins. The part that is opposite controls the orientation of the motion relative to the x axis. It is convenient to separate out these two parts. We will follow the pattern of Equations (3.14) and (3.15). The average initial phase is denoted by $\Sigma/2$:

$$\frac{\Sigma}{2} = \frac{\delta_L + \delta_R}{2}. \quad (3.103)$$

Half the difference is denoted by $\Delta/2$:

$$\frac{\Delta}{2} = \frac{\delta_L - \delta_R}{2}. \quad (3.104)$$

Adding and subtracting these equations,

$$\delta_L = \frac{\Sigma + \Delta}{2}, \quad (3.105)$$

$$\delta_R = \frac{\Sigma - \Delta}{2}. \quad (3.106)$$

Putting Equations (3.105) and (3.106) in our general combination of two opposite rotations, Equation (3.95):

$$\begin{aligned} A &= A_L e^{i(2\pi ft+(\Sigma+\Delta)/2)} + A_R e^{-i(2\pi ft+(\Sigma-\Delta)/2)} \\ &= A_L e^{i(2\pi ft+\Sigma/2)} e^{i\Delta/2} + A_R e^{-i(2\pi ft+\Sigma/2)} e^{i\Delta/2} \\ &= \left[A_L e^{i(2\pi ft+\Sigma/2)} + A_R e^{-i(2\pi ft+\Sigma/2)}\right] e^{i\Delta/2}. \end{aligned} \quad (3.107)$$

3.2 Oscillations in two dimensions

What is the meaning of this? The term $e^{i\beta}$ has the effect of rotating through the angle β. In this case, the contents of the square brackets are rotated through $\Delta/2$. And what is in the square brackets?

$$[A_L e^{i(2\pi ft+\Sigma/2)} + A_R e^{-i(2\pi ft+\Sigma/2)}]$$
$$= A_L \cos(2\pi ft + \Sigma/2) + iA_L \sin(2\pi ft + \Sigma/2) + A_R \cos(2\pi ft + \Sigma/2) - iA_R \sin(2\pi ft+\Sigma/2)$$
$$= (A_L + A_R)\cos(2\pi ft + \Sigma/2) + i(A_L - A_R)\sin(2\pi ft + \Sigma/2). \tag{3.108}$$

Identifying the real and imaginary parts with the x and y components,

$$x = (A_L + A_R)\cos(2\pi ft + \Sigma/2), \tag{3.109}$$

$$y = (A_L - A_R)\sin(2\pi ft + \Sigma/2). \tag{3.110}$$

This represents the ellipse

$$\left(\frac{x}{a}\right)^2 + \left(\frac{y}{b}\right)^2 = 1, \tag{3.111}$$

of semimajor axis

$$a = A_L + A_R \tag{3.112}$$

and semiminor axis

$$b = |A_L - A_R|. \tag{3.113}$$

By varying A_L and A_R, we control a and b. We also control the direction of the motion; with $A_L > A_R$ the rotation is to the left, rotation to the right requires $A_R > A_L$. By varying δ_L and δ_R we control Δ, which allows us to set the angle of the ellipse relative to the x axis. In this way we can construct an ellipse of any semimajor and semiminor axes, at any inclination to the x axis, with motion in either direction, as shown in Figure 3.8.

This approach is much more direct than trying to set those parameters by employing Cartesian coordinates, Equations (3.66), (3.67) and (3.68).

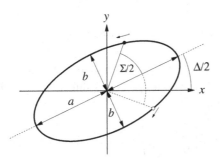

Figure 3.8 Combining two circular oscillations of the same frequency and the different amplitudes (A_L and A_R) and initial phases (δ_L and δ_R). Here the length of the semimajor axis, a, is twice the length of the semiminor axis length, b. This means either $A_L = 3A_R$ (anticlockwise motion, open dot and full lines) or $A_R = 3A_L$ (clockwise motion, open dot and dashed lines). The inclination of the semimajor axis from the x axis is set by $\Delta/2$, where $\Delta = \delta_L - \delta_R$. The starting angle relative to the semimajor axis is set by $\Sigma/2$, where $\Sigma = \delta_L + \delta_R$.

3.3 Jones notation

An ingenious way of describing polarisation is to use the exponential notation of Equation (2.46) in the matrix representation of Equation (3.82):

$$\begin{bmatrix} x \\ y \end{bmatrix} = \begin{bmatrix} A_x e^{i\phi_x} \\ A_y e^{i\phi_y} \end{bmatrix}. \qquad (3.114)$$

We will assume the frequencies are the same in both x and y directions. We further assume that the initial x phase is zero ($\delta_x = 0$) and that the phase difference between the y and x phase is Δ; then $\delta_y = \Delta$. The x and y phases are then:

$$\phi_x = 2\pi f t, \qquad (3.115)$$

$$\phi_y = 2\pi f t + \Delta = \phi_x + \Delta. \qquad (3.116)$$

We are all set up to take the common term $\phi_x = 2\pi f t$ out of the two exponentials,

$$\begin{bmatrix} x \\ y \end{bmatrix} = \begin{bmatrix} A_x e^{i\phi_x} \\ A_y e^{i(\phi_x + \Delta)} \end{bmatrix} = \begin{bmatrix} A_x e^{i\phi_x} \\ A_y e^{i\phi_x} e^{i\Delta} \end{bmatrix} = e^{i\phi_x} \begin{bmatrix} A_x \\ A_y e^{i\Delta} \end{bmatrix}. \qquad (3.117)$$

The magnitude of the final column vector is $\sqrt{A_x^2 + A_y^2}$. We bring this term out to obtain a vector of unit magnitude:

$$\begin{bmatrix} x \\ y \end{bmatrix} = e^{i\phi_x} \begin{bmatrix} A_x \\ A_y e^{i\Delta} \end{bmatrix} = e^{i\phi_x} \sqrt{A_x^2 + A_y^2} \begin{bmatrix} \dfrac{A_x}{\sqrt{A_x^2 + A_y^2}} \\ \dfrac{A_y}{\sqrt{A_x^2 + A_y^2}} e^{i\Delta} \end{bmatrix}. \qquad (3.118)$$

We replace the cumbersome $A_x/\sqrt{A_x^2 + A_y^2}$ with a_x and $A_y/\sqrt{A_x^2 + A_y^2}$ with a_y:

$$\begin{bmatrix} x \\ y \end{bmatrix} = e^{i\phi_x} \sqrt{A_x^2 + A_y^2} \begin{bmatrix} a_x \\ a_y e^{i\Delta} \end{bmatrix}. \qquad (3.119)$$

The vector on the right-hand side contains the essential information we want concerning the relationship between the x and y components. We define this as the **Jones vector**:

$$a \equiv \begin{bmatrix} a_x \\ a_y e^{i\Delta} \end{bmatrix}. \qquad (3.120)$$

In the way we have constructed it, the Jones vector is normalised; that is, its magnitude, $\sqrt{a_x^2 + a_y^2}$, is equal to one.

Let's see how easy it is to express different polarisation states using the Jones vector. Let's start with our prototypical oscillation, purely in the x direction. For this, $A_y = 0$ and so $a_y = 0$. The Jones vector becomes

$$a = \begin{bmatrix} a_x \\ 0 \end{bmatrix} = \begin{bmatrix} 1 \\ 0 \end{bmatrix}. \qquad (3.121)$$

3.3 Jones notation

(We know that a_x must be 1, since we have defined the Jones vector to have a magnitude 1.) On the other hand,

$$a = \begin{bmatrix} 0 \\ 1 \end{bmatrix}, \qquad (3.122)$$

with $a_x = 0$, represents an oscillation linearly polarised along the y axis.

Let's move up to the next level of sophistication and consider the case when the x and y amplitudes are equal (the subject of Section 3.2.1). Then $a_x = a_y$. Since the Jones vector is normalised we can be more specific and deduce $a_x = a_y = 1/\sqrt{2}$. For equal amplitude perpendicular oscillations, the Jones vector is

$$a = \frac{1}{\sqrt{2}} \begin{bmatrix} 1 \\ e^{i\Delta} \end{bmatrix}. \qquad (3.123)$$

Substituting successively into Equation (3.123) the values of $\Delta = 0, \pi/2, \pi, 3\pi/2$ we have for the term $e^{i\Delta}$ successively $1, i, -1, -i$ and the corresponding Jones vectors

$$\frac{1}{\sqrt{2}} \begin{bmatrix} 1 \\ 1 \end{bmatrix}, \quad \frac{1}{\sqrt{2}} \begin{bmatrix} 1 \\ i \end{bmatrix}, \quad \frac{1}{\sqrt{2}} \begin{bmatrix} 1 \\ -1 \end{bmatrix}, \quad \frac{1}{\sqrt{2}} \begin{bmatrix} 1 \\ -i \end{bmatrix}. \qquad (3.124)$$

As in Figure 3.3, these correspond in turn to linear polarisation along the line $x = y$, right circular polarisation, linear polarisation along the line $x = -y$, left circular polarisation. (Some authors swap the left and right Jones vectors given here.)

The phase difference is not restricted to these special values but can take on any value. For example, for $\Delta = \pi/3$ (as depicted in Figure 3.4), the Jones vector is

$$a = \frac{1}{\sqrt{2}} \begin{bmatrix} 1 \\ e^{i\pi/3} \end{bmatrix} = \frac{1}{\sqrt{2}} \begin{bmatrix} 1 \\ \cos \pi/3 + i \sin \pi/3 \end{bmatrix} = \frac{1}{\sqrt{2}} \begin{bmatrix} 1 \\ 1/2 + i\sqrt{3}/2 \end{bmatrix}. \qquad (3.125)$$

The utility of the Jones notation extends to the case when the amplitudes of the oscillations are unequal (the topic of Section 3.2.2). We use the general expression of Equation (3.120). For example, in the case depicted in Figure 3.5, $A_x = 1.5 \times A_y$, so $a_x = 1.5/\sqrt{1.5^2 + 1^2} \sim 0.83$ and $a_y = 1/\sqrt{1.5^2 + 1^2} \sim 0.55$; also, $\Delta = \pi/3$. So the Jones matrix representing the elliptical polarisation of Figure 3.5 is

$$a \sim \begin{bmatrix} 0.83 \\ 0.55 e^{i\pi/3} \end{bmatrix} \sim \begin{bmatrix} 0.83 \\ 0.55(\cos \pi/3 + i \sin \pi/3) \end{bmatrix} \sim \begin{bmatrix} 0.83 \\ 0.28 + 0.48i \end{bmatrix}. \qquad (3.126)$$

We can readily express the gist of Section 3.2.3 within the Jones notation. Adding right and left circularly polarised oscillations results in an oscillation polarised along the x direction:

$$\frac{1}{\sqrt{2}} \begin{bmatrix} 1 \\ i \end{bmatrix} + \frac{1}{\sqrt{2}} \begin{bmatrix} 1 \\ -i \end{bmatrix} = \frac{2}{\sqrt{2}} \begin{bmatrix} 1 \\ 0 \end{bmatrix}. \qquad (3.127)$$

Subtracting left from right circularly polarised oscillations gives an oscillation along the y axis:

$$\frac{1}{\sqrt{2}} \begin{bmatrix} 1 \\ i \end{bmatrix} - \frac{1}{\sqrt{2}} \begin{bmatrix} 1 \\ -i \end{bmatrix} = \frac{2}{\sqrt{2}i} \begin{bmatrix} 0 \\ 1 \end{bmatrix}. \qquad (3.128)$$

We can go further and use the Jones formalism to deal with mechanisms that convert one sort of polarisation into another. For example, let us consider a device that takes an unpolarised oscillation and converts it into an oscillation polarised along the x axis. We could represent this

$$\begin{bmatrix} a_x \\ a_y \end{bmatrix} \rightarrow \begin{bmatrix} 1 \\ 0 \end{bmatrix}. \tag{3.129}$$

The formal, general way to express the linear connection between two column vectors is by a square matrix. It supposes the two components b_x and b_y of the final vector are related by a linear combination of the two components a_x and a_y of the initial vector,

$$b_x = a_{11}a_x + a_{12}a_y,$$
$$b_y = a_{21}a_x + a_{22}a_y. \tag{3.130}$$

This is written compactly as

$$\begin{bmatrix} b_x \\ b_y \end{bmatrix} = \begin{bmatrix} a_{11} & a_{12} \\ a_{21} & a_{22} \end{bmatrix} \begin{bmatrix} a_x \\ a_y \end{bmatrix}. \tag{3.131}$$

The central 2 × 2 matrix in this equation, linking the initial and final Jones vectors, is called the **Jones matrix**. The Jones notation proves useful in the discussion of terahertz components in Chapters 8 and 9. Table 3.1 displays some basic Jones vectors and Table 3.2 displays some Jones matrices.

Table 3.1 Jones vectors

Linear polarisation							
$\leftrightarrow$	$\begin{bmatrix} 1 \\ 0 \end{bmatrix}$	x		$\updownarrow$	$\begin{bmatrix} 0 \\ 1 \end{bmatrix}$		$y\ (= x + \pi/2)$
$\nearrow$	$\frac{1}{\sqrt{2}}\begin{bmatrix} 1 \\ 1 \end{bmatrix}$	$x + \pi/4$		$\searrow$	$\frac{1}{\sqrt{2}}\begin{bmatrix} 1 \\ -1 \end{bmatrix}$		$x - \pi/4$
Circular polarisation							
$\circlearrowright$	$\frac{1}{\sqrt{2}}\begin{bmatrix} 1 \\ i \end{bmatrix}$	right		$\circlearrowleft$	$\frac{1}{\sqrt{2}}\begin{bmatrix} 1 \\ -i \end{bmatrix}$		left

Example 3.2 Quarter-wave retarder

(a) What is the effect of a quarter-wave retarder (fast axis x) on an oscillation polarised along a line inclined at $\pi/4$ to the x axis? (b) On a left circular polarised oscillation?

Solution

From Table 3.2 we see that the matrix for the quarter-wave retarder is $e^{i\pi/4}\begin{bmatrix} 1 & 0 \\ 0 & -i \end{bmatrix}$.

From Table 3.1 we see that the vector for the specified linearly polarised oscillation is

3.3 Jones notation

Table 3.2 Jones matrices

Linear polariser

$\begin{bmatrix} 1 & 0 \\ 0 & 0 \end{bmatrix}$ x $\begin{bmatrix} 0 & 0 \\ 0 & 1 \end{bmatrix}$ $y\,(= x + \pi/2)$

$\dfrac{1}{2}\begin{bmatrix} 1 & 1 \\ 1 & 1 \end{bmatrix}$ $x + \pi/4$ $\dfrac{1}{2}\begin{bmatrix} 1 & -1 \\ -1 & 1 \end{bmatrix}$ $x - \pi/4$

Circular polariser

$\dfrac{1}{2}\begin{bmatrix} 1 & -i \\ i & 1 \end{bmatrix}$ right $\dfrac{1}{2}\begin{bmatrix} 1 & i \\ -i & 1 \end{bmatrix}$ left

$\lambda/4$ retarder

$e^{i\pi/4}\begin{bmatrix} 1 & 0 \\ 0 & -i \end{bmatrix}$ x fast axis $e^{i\pi/4}\begin{bmatrix} 1 & 0 \\ 0 & i \end{bmatrix}$ y fast axis

---

$\dfrac{1}{\sqrt{2}}\begin{bmatrix} 1 \\ 1 \end{bmatrix}$. The result of applying the retarder to the oscillation is

$$e^{i\pi/4}\begin{bmatrix} 1 & 0 \\ 0 & -i \end{bmatrix} \dfrac{1}{\sqrt{2}}\begin{bmatrix} 1 \\ 1 \end{bmatrix} = \dfrac{e^{i\pi/4}}{\sqrt{2}}\begin{bmatrix} 1 \\ -i \end{bmatrix}. \tag{3.132}$$

Consulting Table 3.1, we see that the result corresponds to a left circular polarised oscillation.

(b) Now feeding a left circular polarised oscillation through the retarder,

$$e^{i\pi/4}\begin{bmatrix} 1 & 0 \\ 0 & -i \end{bmatrix} \dfrac{1}{\sqrt{2}}\begin{bmatrix} 1 \\ -i \end{bmatrix} = \dfrac{e^{i\pi/4}}{\sqrt{2}}\begin{bmatrix} 1 \\ -1 \end{bmatrix}. \tag{3.133}$$

The result is an oscillation at angle $-\pi/4$ with respect to the x axis.

The quarter-wave retarder can thus be used to convert linear polarisation to circular polarisation and vice versa.

Exercise 3.10 Linear polariser – Jones notation
Use the Jones notation to demonstrate that a linear polariser parallel to the x direction operating on (a) an oscillation parallel to the x axis, (b) an oscillation at $\pi/4$ to the x axis, (c) a right circular oscillation, (d) an oscillation parallel to the y axis, produces an oscillation parallel to the x axis for (a), (b), (c) and nothing for (d). ∎

Exercise 3.11 Circular polariser – Jones notation
Use the Jones notation to show that a right circular polariser operating on a right circular oscillation or an oscillation along the x axis results in a right circular oscillation, but extinguishes a left circular oscillation. ∎

Exercise 3.12 Half-wave retarder – Jones notation

(a) Multiply the Jones matrices for two quarter-wave retarders to obtain the Jones matrix for a half-wave retarder. (b) Show that a half-wave retarder acting on an oscillation inclined at $\pi/4$ to the x axis produces an oscillation inclined at $-\pi/4$ to the x axis, and vice versa. (c) Show that a half-wave retarder acting on a right circular oscillation produces a left circular oscillation and vice versa. (d) (*More difficult*) A half-wave retarder acts on an oscillation polarised along a line making an angle of α with the x axis. Show that the resulting oscillation is polarised along a line making an angle of $-\alpha$ with the x axis. ■

3.4 Summary

3.4.1 Key terms

principle of superposition, 26   interference, 30   polarisation, 38
beats, 31

3.4.2 Key equations

interference $A_1 + A_2 = 2A_0 \cos\left(\dfrac{\delta_1 - \delta_2}{2}\right) \cos\left(2\pi f t + \dfrac{\delta_1 + \delta_2}{2}\right)$ (3.22)

beats $A_1 + A_2 = 2A_0 \cos[2\pi \dfrac{f_1 - f_2}{2} t] \cos[2\pi \dfrac{f_1 + f_2}{2} t + \delta]$ (3.29)

3.5 Table of symbols, Chapter 3

General mathematical symbols appear in Appendix B. If the unit of a quantity depends on the context, this is denoted '—'.

Symbol	Meaning	Unit
a	Jones vector	—
a_x, a_y	Jones vector component	—
a_{ij}	Jones matrix element	—
A_0, A_0', A_0''	amplitude of oscillation	—
A_1, A_2	amplitude of first, second oscillation	—
A_L, A_R	amplitude of left, right oscillation	—
A_x, A_y	amplitude of x, y oscillation	—
b_x, b_y	Jones vector component	—
f, f', f''	frequency	Hz
f_0	frequency, fundamental	Hz
r	modulus of complex number	—
t	time	Hz^{-1}
x	coordinate label	—
$x_{\circlearrowleft}$	coordinate label for left rotation	—
$x_{\circlearrowright}$	coordinate label for right rotation	—
$\hat{x}$	unit vector in x direction	—
$\hat{X}$	unit vector in rotated x direction	—
y	coordinate label	—
$y_{\circlearrowleft}$	coordinate label for left rotation	—
$y_{\circlearrowright}$	coordinate label for right rotation	—
$\hat{y}$	unit vector in y direction	—
$\hat{Y}$	unit vector in rotated y direction	—
$\tilde{z}$	complex quantity	—
α	general angle	rad
β	general angle	rad
$\delta, \delta', \delta''$	initial phase	rad
δ_1, δ_2	initial phase of first, second oscillation	rad
δ_L, δ_R	initial phase of left, right oscillation	rad
δ_x, δ_y	initial phase of x, y oscillation	rad
Δ	difference of initial phases	rad
θ	general angle	rad
θ	argument of complex number	rad
ϕ	phase	rad
ϕ_1, ϕ_2	phase of first, second oscillation	rad
Σ	sum of initial phases	rad
ω	angular frequency	rad Hz

4 LIGHT

This chapter calls on maths, but the maths is relatively elementary. Section 4.1 requires only the basic operations of addition, subtraction, multiplication and division. Section 4.2 requires calculus. Section 4.3 requires trigonometry, but is a rather direct extension of the methods we have already met in Chapter 2. Section 4.5 assumes some geometrical facility in picturing planes, circles and ellipses. Section 4.6 assumes a background in integration and differentiation.

Let there be light! Light is a basic physical entity. One might argue that light is the most fundamental physical phenomenon of all.

Physics is about energy and matter and their interaction. This chapter is about energy in its purest form, energy in the form of light. Chapter 5 is about matter. Chapter 6 is about the interaction between light and matter. These three chapters constitute the physics core of the book.

We start with the *photon* (Section 4.1). A photon is a bunch, a packet, a particle of light. We will look at what it means to describe light in this way.

An alternative way to think about light is to think of it as a *wave*. Maxwell's equations lead inexorably to the conclusion that light is an *electromagnetic wave* (Section 4.2).

Waves are a generalisation of oscillations. Oscillations involve only one variable, time. Waves add to this an additional variable, displacement. So waves can be described mathematically by extending the ideas of Chapter 2 in a straightforward manner. This extension from oscillation to wave is the subject of Section 4.3.

Waves travelling through space may have different geometries. Loosely speaking, waves may have different shapes. The technical way to describe this is by the *polarisation* of the wave. In Section 4.5, we meet *linearly polarised*, *circularly polarised* and *elliptically polarised* waves.

Thus far, we have talked about light without referring at all to matter. Our discussion so far has been about light in free space, or the vacuum. We conclude the chapter by our first foray into matter, as a segue into the next chapter. We consider a fundamental way by which any piece of matter at all may produce light. We will see that any heated object, that is, any object at a temperature above absolute zero, will radiate light. Such objects include you, me and the lamppost, before we even get as far as the lamp. For simplicity, we look at the ideal form of radiation from a hot object, in Section 4.6, on *blackbody radiation*.

Learning goals

By the time you finish this chapter you should be able to describe:

- light in terms of photons,
- light in terms of waves,
- waves in mathematical terms,
- blackbody radiation.

4.1 Photons

The energy of a **photon** is directly proportional to its frequency. No deeper statement this book makes.

$$E = hf = \hbar\omega. \tag{4.1}$$

The symbol h stands for the *Planck constant*. The symbol $\hbar$, pronounced 'h-bar', stands for the Planck constant divided by 2π. It is also called *the reduced Planck constant* and *the natural unit of action*. In SI units,

$$h = 6.626\,069\,57(29) \times 10^{-34} \text{ J s} \tag{4.2}$$

and

$$\hbar = 1.054\,571\,726(47) \times 10^{-34} \text{ J s}. \tag{4.3}$$

The numbers in brackets indicate the uncertainty in the final two digits.

Energy, E, is measured in joules (J) in the SI system. So the units of h are joules per hertz or joules-seconds. The joule, in turn, is equivalent to the newton-metre (N m). The newton (N) is the unit of force and is expressed in terms of the SI base units as kg m s^{-2}. Here I have introduced for the first time a new SI base unit, the kilogram (kg), the base unit for the base quantity mass. Energy may also be expressed in terms of electrical, rather than mechanical, quantities. To do this I introduce the SI base quantity electric current. Electric current has the base unit of the ampere (A). Electric charge then has the units of ampere-seconds (A s), equivalent to the derived unit of the coulomb (C). Electric potential in turn has units of the volt (V), equivalent to the base unit of kg m^2 s^{-2} C^{-1}. From this perspective, energy may be expressed in terms of electron volts (eV), a non-SI unit, but accepted for use with the SI. Some photon energies in the terahertz range are given in Table 4.1.

Table 4.1 Energies of terahertz photons

Photon frequency, f	0.1 THz	1.0 THz	10 THz
Energy, E	6.626×10^{-23} J	6.626×10^{-22} J	6.626×10^{-21} J
Energy, E	0.414 meV	4.14 meV	41.4 meV

Rather than use the SI system, we might construct a system of units in which h was set to the value of one. Then we would write Equation (4.1) as $E = f$. This is a profound statement indeed; energy and frequency amount to the same thing.

Taking differences of each side of Equation (4.1), $\Delta E = h\Delta f$, multiplying both sides by Δt, and then invoking the time-bandwidth theorem (Equation 2.47), $\Delta f \Delta t \geq 1$, leads to

$$\Delta E \Delta t \geq h. \qquad (4.4)$$

The longer a photon exists, the more precisely its energy may be known.

The photon has momentum. The direction of the momentum is the same as the direction the photon is travelling. The magnitude of the momentum, p, is

$$p = \frac{h}{\lambda} = h\sigma = \hbar k. \qquad (4.5)$$

We met the wavelength, λ, and the wavenumber, σ, in Chapter 1. A new quantity is k, the magnitude of the angular wavenumber. I will have more to say about this in Section 4.3, but will mention in passing that $k = 2\pi\sigma = 2\pi/\lambda$. Some photon wavelengths and wavenumbers in the terahertz range are given in Table 4.2.

Table 4.2 Wavelength and wavenumber of terahertz photons

Photon frequency, f	0.1 THz	1.0 THz	10 THz
Wavelength, λ	3 mm	300 µm	30 µm
Wavenumber, σ	3.34 cm^{-1}	33.4 cm^{-1}	334 cm^{-1}

In common usage, 'light' refers to photons that the eye can see. I will refer to these photons as 'visible light'. In this book, and in physics generally, 'light' is taken to refer more broadly to photons of any frequency. This physics usage is based on the recognition that there is no fundamental difference between photons the eye can see and photons the eye cannot see.

Visible light is usually taken to correspond to photons with frequencies in the range from a little over 400 THz (blue light) to a little under 800 THz (red light). This is a ratio of about two in frequency (an octave). In this book, the terahertz range is taken to be 0.1 to 10 THz. This is a ratio of 100 (two decades). So the terahertz range is 50 times greater than the visible range! Be aware of the huge range of frequency ratios encompassed by 'terahertz' compared to 'visible' light.

4.2 Electromagnetic waves

Maxwell's equations bring together electricity and magnetism. They show electricity and magnetism are intimately related. Two sides of the one coin, one might say. The marriage of electricity and magnetism is electromagnetism. What is more, the solution of Maxwell's equations yields an electric field coupled with a magnetic field propagating through space. The coupled, propagating electric and magnetic fields are called an **electromagnetic wave** – light, by another name.

4.2 Electromagnetic waves

I expect you know something about electricity and about magnetism already, but, in case you do not, I will very briefly introduce here the key ideas we need in order to go on. The electric field, denoted **E**, has units of volts per metre (or newtons per coulomb). The magnetic field, denoted **B**, has units of tesla, equivalent to kg s^{-1} C^{-1}. Both **E** and **B** are vectors, which is indicated by them being written in bold font. Regarding the mathematics we need, $\nabla \cdot$ is the divergence operator and $\nabla \times$ is the curl operator. Both relate to differentiation in space and so both introduce a unit of inverse metre.

In the progress of this book, we are dealing with light before we deal with matter. So the statement of Maxwell's equations here has to do with light in the absence of matter. (Specifically, in case you know about such things already, there are no charges and no currents, but I have not introduced the concept of charge or current yet.) Here are Maxwell's four equations for light in empty space:

$$\nabla \cdot \mathbf{E} = 0. \tag{4.6}$$

$$\nabla \cdot \mathbf{B} = 0. \tag{4.7}$$

$$\nabla \times \mathbf{E} = -\frac{\partial \mathbf{B}}{\partial t}. \tag{4.8}$$

$$\nabla \times \mathbf{B} = \mu_0 \epsilon_0 \frac{\partial \mathbf{E}}{\partial t}. \tag{4.9}$$

Equation (4.6) says there are no electric charges in empty space. Equation (4.7) says there are no magnetic charges in empty space. (Are there magnetic charges, or monopoles, anywhere? That is another story.) Equation (4.8) and Equation (4.9) are similar in form. Each says the curl of one field is directly related to the partial derivative with respect to time ($\partial/\partial t$, Appendix B) of the other field. Two sides of the one coin. The symbols ϵ_0 and μ_0 stand for fundamental physical constants related to electricity and magnetism.

The electric constant

$$\epsilon_0 = 8.854\,187\,817 \cdots \times 10^{-12} \text{ F m}^{-1}. \tag{4.10}$$

The magnetic constant

$$\mu_0 = 4\pi \times 10^{-7} = 12.566\,370\,614 \cdots \times 10^{-7} \text{ N A}^{-2}. \tag{4.11}$$

No uncertainties are given for the electric and magnetic constants as the values are exact (for reasons I will explain later). I introduced a moment ago the units of newtons (N) and amperes (A) that appear in the units of the magnetic constant. The unit of the farad (F) appears in the units of the electric constant. The farad may be expressed as a coulomb per volt (C V^{-1}).

Now let us look more closely at the set of four equations and see where it leads. In doing so, we will retrace the footsteps of Maxwell, and see that a solution of the equations is a set of propagating electromagnetic waves.

Let us first consider the form of Maxwell's equations and the relationships between them before formally deriving the required equations. Equation (4.8) relates a differentiation in space (on the left-hand side), with a differentiation in time (on the right-hand side). But the quantities differentiated are different – the electric field (on the left-hand side) and the magnetic field (on the right-hand side). So we see the electric and magnetic fields are coupled together in a most interesting way: the spatial derivative of the electric field is related to the temporal derivative of the magnetic field. In some sense, space and time are being treated on an equal footing. A change with time of one field is echoed by a change with space of the other field. An (albeit poor) picture might be a couple dancing, not in one spot, but twirling down the dance hall, their motion in time with the music, tracing a sinuous path as they move, the motion of the first influencing the second, and the motion of the second influencing the first.

How can we untangle this? Well, we can't, that's the point: the electric and magnetic fields are irrevocably intertwined. But we can solve Equation (4.8) – not by itself, but by invoking Equation (4.9). Equation (4.9) is something of a complement to Equation (4.8). Equation (4.9) says the spatial derivative of the magnetic field is related to the temporal derivative of the electric field. So here is the strategy. Start with, say, the magnetic field. We can relate the time derivative of the magnetic field to the electric field (Equation 4.8), then relate the time derivative of that to the magnetic field (Equation 4.9). Going through the ringer twice, we can get rid of the electric field. Or, we could have started with the electric field and removed the magnetic field. Taken together, Maxwell's equations let us eliminate either the electric field or the magnetic field to have an equation involving just one or the other.

Let's follow that wordy description with the mathematical solution of Maxwell's equations. The derivation assumes some familiarity with vector manipulation. The steps are given in the exercise below, if you wish to follow them in detail. The result is

$$\frac{\partial^2 \mathbf{E}}{\partial t^2} - \frac{1}{\mu_0 \epsilon_0} \nabla^2 \mathbf{E} = 0 \qquad (4.12)$$

and

$$\frac{\partial^2 \mathbf{B}}{\partial t^2} - \frac{1}{\mu_0 \epsilon_0} \nabla^2 \mathbf{B} = 0. \qquad (4.13)$$

You might notice Equation (4.12) and Equation (4.13) are precisely the same in form. Replacing $\mathbf{E}$ in Equation (4.12) with $\mathbf{B}$ gives Equation (4.13). The near-symmetry of the electric and magnetic fields in Maxwell's equations (Equations 4.6–4.9) leads to a perfect symmetry in the solution (Equations 4.12 and 4.13). I will show in the next section that this form of equation is an equation for a wave and that the factor in front of the ∇^2 term is the square of the speed of the wave. Introducing the symbol c to represent the speed of light in a vacuum,

$$c = \frac{1}{\sqrt{\mu_0 \epsilon_0}} = 299\,792\,458 \text{ m s}^{-1} = 299\,792\,458 \text{ m Hz}. \qquad (4.14)$$

Let me say a few words about this key equation. If you have studied relativity, you may know that the speed of light in a vacuum is independent of the speed of the origin (the source) or the destination (the receiver). The speed of light in a vacuum is strongly regarded as being constant by physicists. Could we call it the most 'constant' of 'constants'? A consequence of this high view physicists have of the constancy of c is that the value for c is fixed in the SI system. The speed of light cannot be measured within the SI system; it is defined to be 299 792 458 m s^{-1}. If you are going to commit to memory any constant in physics, why not this one? (It is not that difficult. It is probably shorter than your phone number. You might adopt my preferred unit as well, the metre-hertz.)

We can now briefly revisit the values of μ_0 and ϵ_0. The magnetic constant (Equation 4.11) is defined to be $\mu_0 = 4\pi \times 10^{-7}$ NA^{-2}. This number is exact (albeit irrational). Now, with c and μ_0 exactly defined, the exact (but irrational) value of ϵ_0 follows directly from Equation (4.14), $\epsilon_0 = 1/\mu_0 c^2$.

Armed with the definition of c (Equation 4.14), we may now recast Equations (4.12) and (4.13) into the form in which they are usually written,

$$\frac{\partial^2 \mathbf{E}}{\partial t^2} - c^2 \nabla^2 \mathbf{E} = 0 \tag{4.15}$$

and

$$\frac{\partial^2 \mathbf{B}}{\partial t^2} - c^2 \nabla^2 \mathbf{B} = 0. \tag{4.16}$$

The next section will deal in detail with the solution to these equations and how to interpret them. But let me emphasise, in finishing this section, that although these two equations are shown here as separate, and, mathematically speaking, we could now solve one equation independently of the other, physically speaking, the two phenomena of electricity and magnetism are inseparably bound in the electromagnetic wave.

Exercise 4.1 Deriving the electromagnetic wave equations from Maxwell's equations

(This exercise assumes knowledge of vector manipulation.) In this exercise, we will work through the steps to obtain the electromagnetic wave equations (Equations 4.12 and 4.13) from Maxwell's equations in empty space (Equations 4.6 to 4.9).

(a) Use the vector triple product to establish the vector identity

$$\nabla \times (\nabla \times \mathbf{V}) = \nabla(\nabla \cdot \mathbf{V}) - \nabla^2 \mathbf{V}, \tag{4.17}$$

where $\mathbf{V}$ is any vector.

(b) Apply this result to $\mathbf{E}$ and $\mathbf{B}$ and use Equations (4.6) and (4.7) to establish $\nabla \times (\nabla \times \mathbf{E}) = -\nabla^2 \mathbf{E}$ and $\nabla \times (\nabla \times \mathbf{B}) = -\nabla^2 \mathbf{B}$.

(c) Take the curl ($\nabla \times$) of Equation (4.8). On the left-hand side, use the result of part (b) to simplify the result. On the right-hand side, assume the orders of differentiation may be swapped, to obtain $-\partial/\partial t (\nabla \times \mathbf{B})$.

(d) In the resulting equation, use Equation (4.9) to eliminate the magnetic field and Equation (4.14) to eliminate μ_0 and ϵ_0 to obtain Equation (4.15).

(e) Follow a similar strategy, starting with taking the curl of Equation (4.9), to obtain Equation (4.16). ∎

4.3 Describing waves

We will now look at the solution to an equation such as (4.15) and (4.16). These are examples of a wave equation. The solution is a mathematical description of a **wave**.

Chapter 2 was concerned with the description of an oscillation. Now we are looking at something similar, but more complicated. I hope you remember the equation for an oscillation (Equation 2.18):

$$A_0 \cos(2\pi f t + \delta) = A_0 \cos \phi. \tag{4.18}$$

This equation involves only one variable, time t. A wave involves a second variable, to do with space. Let us restrict ourselves to one-dimensional space for the moment. Let us call the space variable z. An equation for a wave is a simple extension of the equation for an oscillation:

$$A_0 \cos(kz - 2\pi f t - \delta) = A_0 \cos \phi. \tag{4.19}$$

Here k is a constant. I will explain the meaning of k and the derivation of this equation as we go along; for the moment, I want you to observe that Equation (4.19), describing a wave, may be viewed as a direct extension to Equation (4.18), describing an oscillation. Put the opposite way, an oscillation may be seen as a wave monitored at a single location ($z = 0$).

Let me clarify the signs I have employed in Equation (4.19), because this is sometimes a source of confusion. If we put $z = 0$ into Equation (4.19) we obtain $A_0 \cos(-2\pi f t - \delta)$. This might look different from Equation (4.18), $A_0 \cos(+2\pi f t + \delta)$, but in fact they are the same, since $\cos(-\alpha) = \cos \alpha$ (Appendix C). So the wave, which is defined for any value of z, reduces to an oscillation if we fix z. The equations are such that at the particular value of $z = 0$ the wave we have introduced here becomes the oscillation discussed in Chapters 2 and 3.

Let me now clarify the signs of the z and t terms. Setting aside the initial phase, δ, for a moment, the phase is $kz - 2\pi f t$. Remember that the variables in this expression are z and t and we are taking k and f (and 2 and π) to be constants. The negative signifies that increasing z changes the phase in the opposite sense to increasing t. So to keep the phase the same – to be at the maximum or crest of the wave for example, corresponding to the phase being zero – as t increases, so must z. This means the crest moves to larger values of z as time goes by. In other words, the wave moves in the positive z direction. We could achieve the same result by writing the phase as $2\pi f t - kz$, and some authors do; but $kz - 2\pi f t$ is more standard. Be aware that $kz + 2\pi f t$ means something else, a wave progressing in the opposite or $-z$ direction.

Many of the results for oscillations set out in Chapters 2 and 3 carry over directly to waves. This is because the equation for the wave can be written in exactly the same form as the equation for an oscillation, $A_0 \cos \phi$. The only difference is in the detail of the phase, which is $2\pi f t + \delta$ for the oscillation and $kz - 2\pi f t - \delta$ for the wave.

Let us now look at the solution of an equation such as (4.15) or (4.16). We will look first at the mathematics, then look into the physical meaning. Let's start with Equation (4.15) in only one dimension. With this simplification we can replace the vector **E** with the scalar E, and replace the curl operator with differentiation with respect to z:

$$\frac{\partial^2 E}{\partial t^2} - c^2 \frac{d^2 E}{dz^2} = 0. \tag{4.20}$$

This is an example of the one-dimensional wave equation, which may be written in general as

$$\frac{\partial^2 u}{\partial t^2} = v^2 \frac{d^2 u}{dz^2}, \tag{4.21}$$

where u is a scalar quantity and v is the speed of the wave.

Let's now consider the form of the solution we might expect for a wave equation, Equation (4.20) or (4.21). For a start, the two variables z and t are involved, so we will assume the quantity E or u depends on these variables, that is $u = u(z,t)$. (It may not, but the problem then becomes simpler and less interesting.) We see that differentiating twice with respect to time, t, amounts to differentiating twice with respect to distance, z. (The quantity v also appears, which we take to be constant.) We have to involve distance and time in some way, so let's start with the simplest possibility, of multiplying distance with a factor a to make it dimensionless (a will have units of inverse distance) and multiplying time with a factor b to make it dimensionless (b will have units of inverse time) and add these together, $az+bt$. Let's write $u(z,t) = u(az+bt)$. Then $\partial^2 u/\partial t^2 = b^2 u$ and $d^2 u/dz^2 = a^2 u$; so we see that $v = b/a$.

Thus the wave equation can be solved by a large class of functions; the key point is that the function has an argument of the form $(az + bt)$. In particular, Equation (4.19) is a solution to Equation (4.20). Substituting (4.19) into (4.20) yields

$$c = \frac{2\pi f}{k} = \frac{\omega}{k}. \tag{4.22}$$

Let us spend a moment to consider f, ω, k, c and their relations. Regarding the time variable, f is the frequency and $\omega = 2\pi f$ is the angular frequency associated with an oscillation. Likewise, k may be viewed as characterising the recurrence of an oscillation, but an oscillation not in time, but in space. We may call k the 'period' in space related to the **wavelength**, denoted λ. Just as in time $T = 2\pi/\omega = 1/f$, so in space $\lambda = 2\pi/k = 1/\sigma$, where σ is the **spatial frequency** (also known as the **wavenumber**). Many of the statements about oscillations in Chapter 2 may be carried directly over to the spatial equivalent by making the substitutions $t \Leftrightarrow x$, $T \Leftrightarrow \lambda$, $f \Leftrightarrow \sigma$, $\omega \Leftrightarrow k$. Some parallels are set out in Table 4.3.

The quantity in the phase related to the time variable may be written as

$$2\pi f t \quad \text{or} \quad \frac{2\pi}{T} t \quad \text{or} \quad \omega t. \tag{4.23}$$

The quantity in the phase related to the space variable may be written as

$$2\pi \sigma z \quad \text{or} \quad \frac{2\pi}{\lambda} z \quad \text{or} \quad kz. \tag{4.24}$$

Table 4.3 Quantities used in describing waves

Time variable	Period	Frequency	Angular frequency		
t	T	$f = \dfrac{1}{T}$	$\omega = 2\pi f$	$\omega = \dfrac{2\pi}{T}$	$f = \dfrac{\omega}{2\pi}$
Space variable	Wavelength	Wavenumber	Angular wavenumber		
x	λ	$\sigma = \dfrac{1}{\lambda}$	$k = 2\pi\sigma$	$k = \dfrac{2\pi}{\lambda}$	$\sigma = \dfrac{k}{2\pi}$

4.4 Describing electromagnetic waves

4.4.1 One-dimensional waves

In the previous section, we saw $A_0 \cos(kz - 2\pi ft - \delta) = A_0 \cos\phi$ is a solution to the one-dimensional wave equation $\partial^2 u/\partial t^2 = v^2 d^2 u/dz^2$. Likewise, a solution to $\partial^2 E/\partial t^2 = c^2 d^2 E/dz^2$ (Equation 4.20) is

$$E = E_0 \cos(kz - 2\pi ft - \delta). \quad (4.25)$$

We may also write the solution in the more compact forms

$$E = E_0 \cos\phi = \text{Re}(E_0 e^{i\phi}). \quad (4.26)$$

As I have mentioned before regarding the final form, the notation indicating the real part, Re, is often dropped and the taking of the real part implied. A schematic representation of an electromagnetic wave is given in Figure 4.1.

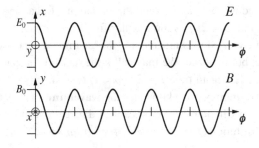

Figure 4.1 An electromagnetic wave. The electric field starts in the x direction (up the page in the top graph and into the page in the bottom graph). The inseparable magnetic field is perpendicular to that, starting in the y direction (out of the page in the top graph and up the page in the bottom graph). The horizontal axis is the phase, $kz - 2\pi ft - \delta$.

At minimum, three dimensions are involved in describing an electromagnetic wave. The electric field is in one direction; the magnetic field is perpendicular to that; the wave propagates perpendicular to both. We usually focus on the electric field, and I have called its direction x. The magnetic field is perpendicular to that; I have called that the y direction. The propagation direction is perpendicular to both, the z direction. If we know the direction of one field and the direction of propagation, we can deduce the direction of the other field. It is usual to specify the electric field and the propagation direction and omit the magnetic field in the description. If the electric field is given as a function of one spatial variable, z in Equation (4.25), we can refer to the description as 'one-dimensional', but bear in mind the electric field is perpendicular to this direction and the magnetic field is in a third direction still. We say light is a *transverse wave* since the fields change in directions perpendicular to the direction of propagation.

In Figure 4.1, the fields are shown as functions of phase, $E_0 \cos\phi$ and $B_0 \cos\phi$. We can view this in different ways. Setting $t = \delta = 0$, for example, gives E and B as a function of distance, z. Alternatively, setting $z = \delta = 0$ gives a plot of E and B against time, t.

4.4.2 Plane waves

Let us now extend this one-dimensional description into three dimensions. Let's assume now that the electric field is a vector with x, y and z components: $\mathbf{E} = (E_x, E_y, E_z)$. What would follow if the vector $\mathbf{E}$ obeyed the same equation as the scalar E,

$$\mathbf{E} = \mathbf{E}_0 \cos(kz - 2\pi ft - \delta)? \tag{4.27}$$

It would mean that, whatever the values of x and y, they do not affect the result. For a given z, the value of $\mathbf{E}$ is the same, regardless of the values of x and y. In three-dimensional space, all the points with a given value of z correspond to a plane cutting the z axis and perpendicular to it. The value of $\mathbf{E}$ is the same everywhere in this plane. For that reason, a wave like this is called a *plane wave*. As time changes, the phase is preserved if z changes; thus the wave moves in the z direction. In the same way, the equation $\mathbf{E} = \mathbf{E}_0 \cos(kx - 2\pi ft - \delta)$ represents a plane wave travelling in the x direction and $\mathbf{E} = \mathbf{E}_0 \cos(ky - 2\pi ft - \delta)$ represents a plane wave travelling in the y direction. More generally, writing the displacement from the origin as $\mathbf{r} = (x, y, z)$, the equation

$$\mathbf{E} = \mathbf{E}_0 \cos(\mathbf{k} \cdot \mathbf{r} - 2\pi ft - \delta) \tag{4.28}$$

is a plane wave travelling in the direction of the vector $\mathbf{k} = (k_x, k_y, k_z)$. Here $\mathbf{k}$ is an example of an *angular wavevector*, the generalisation into two, three or more dimensions of the angular wavenumber k. We won't have to call on this generalisation much in the future, though, since the wave is moving in a single direction, and we simply define one of the Cartesian coordinates, usually z, to be in that direction. Then we can replace $\mathbf{k} \cdot \mathbf{r}$ with kz.

4.4.3 Spherical waves

A completely different geometry is one of spherical symmetry. A function is spherically symmetric if its value only depends on the distance from the origin. In symbols, the function g is spherically symmetric if $g(\mathbf{r}) = g(r)$. I won't prove but will simply state that for a spherically symmetric function,

$$\nabla^2 g = \frac{1}{r} \frac{\partial^2(rg)}{\partial r^2}. \tag{4.29}$$

Substituting this into an equation of the form of Equation (4.15) gives

$$\frac{\partial^2(rg)}{\partial t^2} = c^2 \frac{\partial^2(rg)}{\partial r^2}. \qquad (4.30)$$

From our previous work, we know a solution of this equation is

$$rg = g_0 \cos(kr - 2\pi ft - \delta), \qquad (4.31)$$

or

$$g = \frac{g_0}{r} \cos(kr - 2\pi ft - \delta). \qquad (4.32)$$

We can apply this general result to an electric field. If the electric field is spherically symmetric, that is, if $\mathbf{E}(\mathbf{r}) = \mathbf{E}(r)$, then a solution of Equation (4.15) is

$$\mathbf{E} = \frac{\mathbf{E}_0}{r} \cos(kr - 2\pi ft - \delta). \qquad (4.33)$$

This is the description of a *spherical wave*. As spherical waves move farther and farther from the source, the wavefront becomes flatter and flatter. The flattening wavefront, over which the phase is constant, comes to resemble the plane over which the phase is constant that characterises a plane wave. We will use Equation (4.25) much more often than Equation (4.33).

4.4.4 Standing waves

By direct substitution into Equation (4.20) you should be able to see that not only is $E = E_0 \cos(kz - 2\pi ft)$ a solution, but so is $E = E_0 \cos(kz + 2\pi ft)$, and so is the sum of these,

$$E = E_0 \cos(kz - 2\pi ft) + E_0 \cos(kz + 2\pi ft). \qquad (4.34)$$

Physically, this equation represents two waves of the same frequency and same angular wavenumber but travelling in opposite directions. Using the identity for $\cos\alpha + \cos\beta$ (Appendix C), we can express this as

$$E = 2E_0 \cos(kz) \cos(2\pi ft). \qquad (4.35)$$

At any given point in space (fixed z), the field oscillates as $\cos(2\pi ft)$. At any given point in time (fixed t), the field oscillates with distance as $\cos kz$. At the positions corresponding to $kz = \pi/2, 3\pi/2, 5\pi/2, \ldots$ and at the times corresponding to $2\pi ft = \pi/2, 3\pi/2, 5\pi/2, \ldots$ the field is zero. Precisely between these zeros the field is furthest from zero. This solution to the wave equation is called a *standing* or *stationary* wave.

4.5 Polarisation

The topic of polarisation has been dealt with in the context of oscillations in Sections 3.2.1 to 3.3. The concept is easily extended to waves. Many of the results of Sections 3.2.1 to 3.3 follow directly.

Polarisation, when applied to light, has to do with the orientation of the electric field around the propagation direction. (Once the direction of propagation and the electric field direction are known, the magnetic field direction is set, so there is no need to discuss the magnetic field separately.) I should mention in passing that the word 'polarisation' has other meanings in other contexts (for example, it is applied to the orientation of electric dipoles in a material), so caution should be applied in understanding the word in context.

The polarisation of light will be discussed now for light travelling through a vacuum. That is, the intrinsic features of light's polarisation will be treated. In practice, materials are often used to produce polarised light, or to manipulate the polarisation (for example, to 'flip', or 'rotate' it). Moreover, the way light interacts with materials (for example, how it is reflected) depends on the polarisation. These aspects of polarisation, pertaining to the interaction of light and matter, will be treated in Chapter 5.

Imagine we are looking straight down the path of some light coming towards us (Figure 4.2). We follow the motion of the electric field vector. Perhaps the electric field vector moves back and forth in a straight line (Figure 4.2a). This motion in a line, or linear motion, we call **linear polarisation**. (If we took a three-dimensional view, we would see the electric field vector was always in the same plane, containing that line and perpendicular to the page; for this reason, linearly polarised light is also referred to as having **plane polarisation**.) Perhaps the electric field vector moves around a circle (Figure 4.2b/c). This is known as **circular polarisation**. We may further classify circularly polarised light depending on the sense of the rotation. If the electric field vector rotates clockwise, we refer to **right circular polarisation** (rcp) (Figure 4.2b). If the electric field vector rotates anticlockwise, it is **left circular polarisation** (lcp) (Figure 4.2c). (Let me warn you, some people use exactly the opposite directions!) If the electric field vector traces out an elliptical path, in either direction, we refer to **elliptical polarisation** (Figure 4.2d). Linearly and circularly polarised light may be regarded as special cases of the more general case of elliptically polarised light. Of course, the electric field vector may trace out a combination of these or some other pattern again; these are merely some particular simple possibilities. If the direction of the electric field vector changes randomly with time, we say the light is unpolarised.

4.6 Blackbody radiation

Hot objects give off light. The hotter the object, the more light it gives off.

Calling an object 'hot' might bring to mind the idea of temperature. The hotter the object, the higher the temperature. This is common knowledge.

I expect you have some technical knowledge of temperature from your previous study of thermodynamics. If not, it doesn't matter. I will sum up all you need to know now in a few sentences. Temperature is denoted by the symbol T. It is measured on an absolute scale called the thermodynamic temperature scale. The SI unit for temperature is the kelvin. Unless I specify otherwise, I will always refer to absolute temperature measured in kelvins. The Celsius degree is the same size as the kelvin and so a temperature

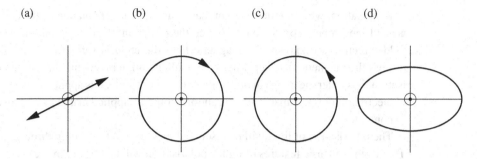

Figure 4.2 Polarisation of light. The light is coming directly out of the page, as indicated by the symbol ⊙. The tip of the electric field vector moves along the paths shown. (The electric field vector is always perpendicular to the direction of light propagation, and so always lies in the page; light is a transverse wave.) (a) Linearly polarised light. The electric field vector always lies on the same straight line. (Also called plane-polarised light.) (b) Circularly polarised light – right circular polarisation (rcp). The tip of the electric field vector rotates clockwise around a circle. (The electric field vector tip traces out a right-hand screw in space.) (c) Circularly polarised light – left circular polarisation (lcp). The tip of the electric field vector rotates anticlockwise around a circle. (The electric field vector traces out a left-hand screw in space.) (d) Elliptically polarised light. The tip of the electric field vector traces an ellipse. The rotation may be in either direction (not shown). Linearly polarised and circularly polarised light may be regarded as special cases of elliptically polarised light.

difference has the same numerical value whether expressed in Celsius degrees or kelvins. The Celsius scale is offset by 273.15 from the kelvin scale; that is, the temperature in degrees Celsius is obtained by subtracting 273.15 from the temperature in kelvin.

The energy associated with the thermodynamic temperature T is given by

$$E = k_B T, \qquad (4.36)$$

where k_B is the Boltzmann constant. This is a rather profound relation, along the lines of Equation (4.1). (Not quite as profound, in my view, but almost.) Equation (4.36) says that the energy in a thermodynamic system is directly proportional to the temperature.

The constant in Equation (4.36) may be given in various units according to convenience.

$$k_B = 1.380\,6488(13) \times 10^{-23} \text{ J K}^{-1} \qquad (4.37)$$
$$= 8.617\,3324(78) \times 10^{-5} \text{ eV K}^{-1} \qquad (4.38)$$
$$= 0.020\,836\,618(19) \text{ THz K}^{-1}. \qquad (4.39)$$

The last form is convenient to use in the terahertz regime. We may invert the last form to obtain

$$1/k_B = 47.992\,434(44) \text{ K THz}^{-1}. \qquad (4.40)$$

So the inverse Boltzmann constant stands pretty close to 48 K per THz. This is a good

4.6 Blackbody radiation

figure to remember (provided you've already mastered c). Photons in the terahertz frequency range, 0.1–10 THz, have equivalent temperatures in the technologically important range from 4.8 K (liquid helium boils at 4.2 K) to 480 K (room temperature is about 300 K), as shown in Table 4.4.

Table 4.4 Equivalent temperatures of terahertz photons

	0.1 THz	1.0 THz	10 THz
Photon frequency, f			
Equivalent temperature, T	4.8 K	48 K	480 K

Blackbody radiation sounds like an oxymoron. How can a 'black' body emit light? Light from darkness! We need to understand that **blackbody** has a particular technical meaning. The name comes from the fact that the blackbody is a perfect absorber of light. That is, light of all frequencies is perfectly captured by a blackbody, none of the light is scattered, reflected, or transmitted. Of course, this is an ideal object that does not exist in practice, but it is useful for theoretical understanding, and materials can be made to approximate the ideal blackbody. A blackbody also perfectly emits radiation. Indeed, defined this way, the only physical parameter that influences the amount of radiation emitted is the temperature of the blackbody.

In describing blackbody radiation, the ratio of photon to thermal energy hf/k_BT appears. The equations are simplified if either $hf/k_BT \gg 1$ or $hf/k_BT \ll 1$. For terahertz photons, the condition $hf/k_BT \gg 1$ (or $hf \gg k_BT$) can be assured if T is less than, say, 1 K. Likewise, for terahertz photons, the condition $hf/k_BT \ll 1$ (or $hf \ll k_BT$) can be assured if T is more than, say, 1000 K. In many practical cases, corresponding to temperatures in the range 4–500 K, $hf/k_BT \sim 1$, and the full equations need to be employed.

We are now in a position to give the expression for the light emitted from a blackbody. This may be expressed in several equivalent ways. We begin with the *Planck function*, which I will not derive but simply state:

$$B_f(T) = \frac{2hf^3}{c^2} \left[\exp\left(\frac{hf}{k_BT}\right) - 1 \right]^{-1}. \quad (4.41)$$

The quantity $B_f(T)$ is known as *surface brightness* or simply *brightness*. It refers to power emitted in the direction normal to the surface, per unit area, per unit solid angle, per unit frequency (phew!). So the units are W m^{-2} Hz^{-1} sr^{-1}.

Let us now examine the Planck function at the extremes of $hf \gg k_BT$ and $hf \ll k_BT$. First, $hf \gg k_BT$. The Planck function is shown in Figure 4.3.

The condition $hf \gg k_BT$ corresponds to the bottom and right of Figure 4.3. In this case, the ratio $hf/k_BT \gg 1$ and so the term $\exp(hf/k_BT)$ is very large, very much larger than 1. Then the term $\exp(hf/k_BT) - 1$ may, to a very good approximation, be replaced with $\exp(hf/k_BT)$. This yields the Wien function:

$$B_f(T) = \frac{2hf^3}{c^2} \exp\left(-\frac{hf}{k_BT}\right). \quad (4.42)$$

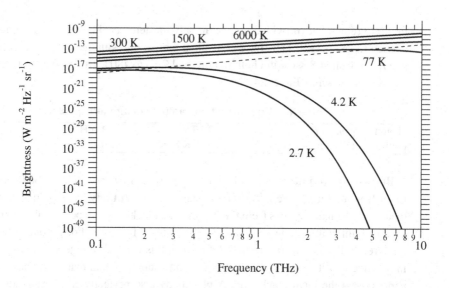

Figure 4.3 Blackbody radiation: the Planck function, illustrating the Wien regime. Logarithmic scales are used on both axes. The frequency range spans 0.1–10 THz. The brightness range spans 40 orders of magnitude! Brightness curves are shown for blackbodies at temperatures 2.7 K (the cosmic background radiation temperature), 4.2 K (the boiling point of liquid helium), 77 K (the boiling point of liquid nitrogen), 300 K (approximately room temperature), 1500 K (approximately the operating temperature of a globar), 6000 K (approximately the surface temperature of the sun). The dashed line joins peak brightness points at differerent temperatures.

As may be seen from Figure 4.3, this is only applicable in the terahertz regime at very low temperatures. So, we will not be referring to this much in future. It is worth noting, however, the brightness in this case decreases with increasing frequency, the $\exp(-f)$ factor dominating the f^3 factor.

Next, $hf \ll k_\text{B} T$.

This corresponds to the top left of Figure 4.3. In this case, the ratio $hf/k_\text{B} T \ll 1$ and so the term $\exp(hf/k_\text{B} T)$ can be written, to a good approximation, as $1 + (hf/k_\text{B} T)$. Then the term $\exp(hf/k_\text{B} T) - 1$ becomes simply $hf/k_\text{B} T$. This yields the Rayleigh-Jeans function:

$$B_f(T) = \frac{2 k_\text{B} T}{c^2} f^2. \tag{4.43}$$

As may be seen from Figure 4.4, this is applicable over the terahertz regime for emitters above room temperature. An object above room temperature is what we usually mean by a 'hot' body. In this case the brightness increases with increasing frequency; specifically, the brightness increases with the square of the frequency. So the brightness at 0.1 THz is not a tenth the brightness at 1 THz, but only one hundredth. This should give you a feel for the fact that blackbodies, even very hot ones, struggle to provide much power in the low-terahertz range.

We have seen at low frequencies ($hf \ll k_\text{B} T$, Equation (4.43)) that brightness increases with frequency and at high frequencies ($hf \gg k_\text{B} T$, Equation (4.42)) that

4.6 Blackbody radiation

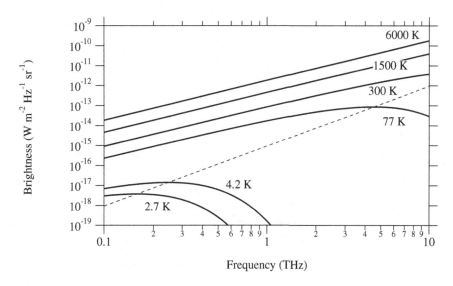

Figure 4.4 Blackbody radiation: the Planck function, illustrating the Rayleigh-Jeans regime. Logarithmic scales are used on both axes. The frequency range spans 0.1–10 THz. The brightness range spans 10 orders of magnitude; this corresponds to the top quarter of Figure 4.3. Brightness curves are shown for blackbodies at temperatures 2.7 K (the cosmic background radiation temperature), 4.2 K (the boiling point of liquid helium), 77 K (the boiling point of liquid nitrogen), 300 K (approximately room temperature), 1500 K (approximately the operating temperature of a globar), 6000 K (approximately the surface temperature of the sun). The dashed line joins peak brightness points at differerent temperatures.

brightness decreases with frequency. Somewhere between these two limits there is a maximum in the Planck function. By differentiating with respect to frequency, you can show the maximum occurs for the frequency f_{peak}

$$f_{peak}[\text{in terahertz}] = 0.058789254(54)T[\text{in kelvin}]. \tag{4.44}$$

So the frequency of the peak of the curve is directly proportional to the temperature. The extremes of the terahertz range, 0.1 and 10 THz, correspond to frequency peaks of blackbodies at 1.7 K and 170 K. These temperatures are well below room temperature. This observation re-inforces what we already know, that 'hot' objects (above room temperature) are on the low side of the blackbody peak, or in the Rayleigh-Jeans region, for radiation at terahertz frequencies. (Equation 4.44 is related to the *Wien displacement law*, which relates to the wavelength (rather than frequency) of the peak in the wavelength (rather than frequency) spectrum.)

Let's consider further 'hot' bodies, that is, ones above room temperature. A practical laboratory source of blackbody radiation is the *globar*, a ceramic rod heated by passing an electrical current through it. Typically, a globar operates at temperatures of about 1500 K. In Figure 4.5, the Planck function is given for objects at temperatures of 500, 1000 and 1500 K, corresponding to about 200, 700 and 1200 degrees Celsius.

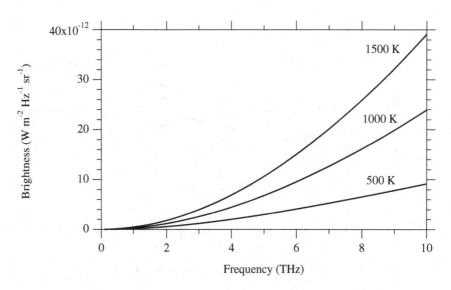

Figure 4.5 Blackbody radiation: the Planck function, illustrating the regime of practical laboratory thermal sources. Linear scales are used on both axes. The frequency range spans 0–10 THz. Brightness curves are shown for blackbodies at temperatures 500, 1000 and 1500 K. In this regime, it is seen the brightness increases approximately quadratically with frequency, $B \propto f^2$, and approximately linearly with temperature, $B \propto T$.

Let's say we are in the Rayleigh-Jeans region of $hf \ll k_B T$ and want to know the total power emitted by a blackbody in a particular frequency range. For example, we might be interested in the frequency range $f_1 = 0.1$ THz to $f_2 = 10$ THz. We simply integrate Equation (4.43) with respect to frequency between the limits of f_1 and f_2 to obtain the power emitted in the direction normal to the surface, per unit area, per unit solid angle, with units W m^{-2} sr^{-1}:

$$P_{\text{R-J}} = \int_{f_1}^{f_2} B_f(T) df = \frac{2k_B}{3c^2}(f_2^3 - f_1^3)T = \sigma_{\text{R-J}} T. \tag{4.45}$$

Let's look at three features of Equation (4.45). First, the dependence on the upper frequency limit is strong, as a cube is involved. For example, if $f_1 = 0.1$ THz and $f_2 = 10$ THz, then $f_2 = 100 f_1$ and f_2^3 is 10^6 times greater than f_1^3. In such a case, to a good approximation, we can set the lower frequency to zero and write $P_{\text{R-J}} = 2k_B T f_2^3/3c^2$. Second, the power emitted is directly proportional to the temperature. Third, this equation is a cut-down version of the Stefan-Boltzmann law. The Stefan-Boltzmann law concerns the radiation over all frequencies ($f_1 = 0$ THz to $f_2 = \infty$ THz) and so necessarily involves the Planck function (Equation 4.41) rather than the Rayleigh-Jeans function (Equation 4.43). (Another difference is that the Stefan-Boltzmann law includes integration over the solid angle of a hemisphere, but this only has the effect of introducing a factor π and reducing the units to W m^{-2}.) The Stefan-Boltzmann law is usually written as $M = \sigma T^4$. The T^4 dependence of the Stefan-Boltzmann law is much stronger than the T dependence of Equation (4.45).

4.6 Blackbody radiation

Exercise 4.2 Planck function – wavelength version

  (This exercise assumes knowledge of calculus.) We may write the Planck function in terms of wavelength rather than frequency via $f = c/\lambda$. Show that

$$B_\lambda(T) = B_f(T)\left|\frac{df}{d\lambda}\right| = \frac{2hc^2}{\lambda^5}\left[\exp\left(\frac{hc}{\lambda k_B T}\right) - 1\right]^{-1}. \tag{4.46}$$

(The units of $B_\lambda(T)$ are W m^{-2} m^{-1} sr^{-1}; that is, power per unit area per unit wavelength per unit solid angle.) ∎

Exercise 4.3 Peak frequency – derivation

  (This exercise assumes knowledge of calculus.) Obtain the peak frequency, Equation (4.44), from the Planck function, Equation (4.41).

  Hint: Differentiate the Planck equation with respect to frequency and set the result to zero to locate the turning point/s. To find the required turning point amounts to solving an equation of the form $(3 - x)e^x = 3$. This has no elementary analytical solution, but may be solved numerically (by trial and error, if you wish). The solution is also given in terms of the Lambert W-function as $x = 3 + W(-3e^{-3})$. Either way, $x = 2.82143\ldots$.
∎

Exercise 4.4 Wien's displacement law

  (It is suggested you complete Exercise 4.2 and Exercise 4.3 before attempting this exercise.)

  (a) Differentiate the wavelength version of the Planck function (Equation 4.46) with respect to wavelength to determine the wavelength, λ_{peak}, at which the function is a maximum.

  Hint: You may encounter an equation of the form $(5 - x)e^x = 5$, which has solution $x = 5 + W(-5e^{-5}) = 4.96511\ldots$

  (b) Show that the product $\lambda_{\text{peak}} T$ is a constant and determine its value in SI units.

  (c) Express Equation (4.44) in terms of wavelength and determine the product of wavelength and temperature in SI units. Why is this different from the answer of part (b)? ∎

Exercise 4.5 The Stefan-Boltzmann law

  (a) Deriving the Stefan-Boltzmann law involves integrating the Planck function (Equation 4.41) over all solid angles; this simply introduces a factor of π and removes the units of sr^{-1} from the units. Now integrate over all frequencies $0 < f < \infty$ to obtain the Stefan-Boltzmann law, $M = \sigma T^4$.

  Hint: You may encounter the definite integral $\int_0^\infty x^3 dx/(e^x - 1)$. This has the value $\pi^4/15$.

  (b) What is the value of σ in terms of fundamental constants?

  (c) What is the value of σ in SI units? ∎

LIGHT

4.7 Summary

4.7.1 Key terms

photon, 55
electromagnetic wave, 56
wave, 60
linear polarisation, 65
circular polarisation, 65
elliptical polarisation, 65
blackbody, 67
blackbody radiation, 67

4.7.2 Key equations

Planck-Einstein relation	$E = hf = \hbar\omega$	(4.1)
general harmonic wave	$A_0 \cos(kz - 2\pi f t - \delta) = A_0 \cos\phi$	(4.19)
Planck function	$B_f(T) = \dfrac{2hf^3}{c^2} \left[\exp\left(\dfrac{hf}{k_B T}\right) - 1 \right]^{-1}$	(4.41)

4.7.3 Additional basic physical quantities

Quantity		Unit		Dimension	
mass	m	kilogram	kg	mass	M
electric current	I	ampere	A	current	I
temperature	T	kelvin	K	temperature	Θ

4.7.4 Derived physical quantities

Quantity		Unit			Dimension
force	F	newton	$N = m\,kg\,s^{-2}$		$F^2\,L\,M$
energy	E	joule	$J = N\,m$		$F^2\,L^2\,M$
charge	q	coulomb	$C = A\,Hz^{-1}$	$= A\,s$	$F^{-1}\,I$
electrical potential	V	volt	$V = J\,C^{-1}$		$F^3\,L^2\,M\,I^{-1}$

4.8 Table of symbols, Chapter 4

General mathematical symbols appear in Appendix B. If the unit of a quantity depends on the context, this is denoted '—'.

Symbol	Meaning	Unit
a	inverse distance	m^{-1}
A	amplitude	—
b	inverse time	Hz
B	magnetic field	$T = V\,s\,m^{-2}$
B_f, B_λ	brightness	$W\,m^{-2}\,Hz^{-1}\,sr^{-1}$
c	lightspeed	m Hz
E	electric field amplitude	$V\,m^{-1} = N\,C^{-1}$
E	energy	$J = N\,m = m^2\,kg\,s^{-1}$
f	frequency	Hz
f_{peak}	frequency, peak	Hz
f_1	frequency, first	Hz
f_2	frequency, second	Hz
h	Planck constant	$J\,Hz^{-1} = J\,s$
$\hbar$	reduced Planck constant	$J\,Hz^{-1} = J\,s$
k	angular wavenumber	$rad\,m^{-1}$
k_B	Boltzmann constant	$J\,K^{-1}$
p	momentum	$kg\,m\,Hz = kg\,m\,s^{-1}$
t	time	Hz^{-1}
T	period	Hz^{-1}
u	scalar quantity	—
v	wavespeed	m Hz
x	coordinate label	—
y	coordinate label	—
z	coordinate label	—
ϵ_0	electric constant	$F\,m^{-1}$
λ	wavelength	m
λ_{peak}	wavelength, peak	m
μ_0	magnetic constant	$N\,A^{-1}$
σ	wavenumber	m^{-1}
ω	angular frequency	rad Hz

5 MATTER

This chapter calls on trigonometry and calculus, both differentiation and integration.

This book is all about things that oscillate at terahertz frequencies, as introduced in Chapter 1. In Chapters 2 and 3 we saw how oscillations can be described using mathematics. In Chapter 4 we moved from the purely mathematical to physical reality, in treating the oscillatory nature of light. Now, in this chapter, we look at oscillations in matter.

The oscillations that can be sustained in matter are described by quantum mechanics. In this chapter, the *wave mechanics* approach to quantum mechanics is introduced via the *Schrödinger equation*. The equation is then solved for many cases of general and practical interest: the free particle, confined states such as in quantum wells and multiple quantum wells, and the oscillations of molecules, both rotational and vibrational. In all cases I deal with here, the emphasis is on the terahertz frequency range.

Learning goals

By the time you finish this chapter you should be able to discuss terahertz phenomena related to

- free particles,
- particles confined in structures such as semiconductor energy bands and quantum wells,
- molecules that are free to rotate and vibrate.

5.1 Wave mechanics

There is no escaping it. We can't really discuss terahertz phenomena in matter without discussing quantum mechanics.

As the name implies, quantum mechanics is about quanta, or distinct amounts. Applied to oscillations, quantum mechanics says that oscillations cannot occur at any arbitrary frequency, but only at particular, distinct frequencies. A lot of this chapter will be dedicated to determining what those frequencies are in particular physical systems.

The species of quantum mechanics we will apply is *wave mechanics*. As the name implies, wave mechanics is based on waves, and directly of utility to us as the frequency of the wave may be determined, and it is (terahertz) frequencies that this book is all about.

Wave mechanics proceeds by writing down the quantum wave equation – the Schrödinger equation – and solving it. How difficult the equation is to solve, and the nature of the solution, depends on the physical system under scrutiny.

Students can find the Schrödinger equation daunting. Fair enough: the equation can only be solved analytically in some cases; in many real cases approximations or numerical solutions must be resorted to. I hope the approach I take here, of easing into the material from simplest to more difficult, will alleviate some of this unease.

5.1.1 Approach

The general strategy will be this:

1. Start by defining a physical system.
2. Write down the Schrödinger equation.
3. Write down the kinetic energy term and write down the potential energy term; these two steps amount to writing down the Hamiltonian.
4. Solve the Schrödinger equation. Usually, I will just tell you the solution. If the solution has an analytical form, you can check for yourself that it works. The solution yields the wavefunction, from which the probability of the particle being at a particular position can be calculated. It also yields the energies of the allowed states.
5. Identify the quantisation condition that is implied in the solution to the Schrödinger equation. This condition often leads to a direct appreciation of the energy quantisation.
6. Express the allowed energy states in terms of equivalent photon frequencies. Likewise, express the energy differences between successive states in terms of frequencies.
7. Identify physical systems that exhibit these phenomena in the terahertz frequency range.

5.2 The Schrödinger equation

The Schrödinger equation appears in many forms. To describe a complicated physical system – for example, one of many dimensions, changing with time, involving relativistic particles – a complicated form is needed. My approach here is to start with the simplest forms of the equation and gradually introduce more complicated versions as they are needed.

The general form of the Schrödinger equation is

$$H\psi = E\psi. \tag{5.1}$$

Let us take a look at each of the three terms that appear in this equation.

ψ is the wavefunction. In general, ψ is a function both of position and of time. The probability of finding a particle at a particular point in time and space is directly related to ψ. (In fact, it is proportional to $\psi^*(\mathbf{r},t)\psi(\mathbf{r},t)$, where $\mathbf{r}$ is the position vector, t is time and * indicates the complex conjugate.)

E is a constant energy. Usually, the equation can be solved for only certain values of E. These are then known as the *allowed energies*, the *eigenenergies*, or simply the *energies* of the system.

H is the Hamiltonian. In classical physics, the Hamiltonian is often simply the total mechanical energy, kinetic plus potential. Then the equation trivially says that the (left-hand side) total energy is equal to the (right-hand side) total energy. In the quantum physics Hamiltonian the energy terms become *operators* rather than algebraic quantities. In the simplest case, in one dimension, the kinetic energy is

$$T = \frac{p^2}{2m} \rightarrow -\frac{\hbar^2}{2m}\frac{d^2}{dx^2}. \tag{5.2}$$

Notice that the classical expression $(p^2/2m)$ is a quantity, while the quantum expression involves not only a quantity $(-\hbar^2/2m)$ but also an operation (d^2/dx^2). The expression for T does not stand alone, but needs to *operate* on the wavefunction.

5.2.1 Time-independent, one-dimensional Schrödinger equation

I mentioned in the preceding section that the wavefunction is in general a function of both time and position. Here we will restrict ourselves to wavefunctions that do not change with time. So time is not a variable in the equation. Wavefunctions that do not change with time are referred to as *stationary states*. (If you find it helpful, you may think of them as *standing waves* that do not propagate.) In general, the wavefunction will exist in three-dimensional space. Here we will restrict ourselves to one dimension. We will label this dimension the x dimension. Rather than write the wavefunction in terms of a three-dimensional position $\mathbf{r}$ and time t, we can simply write it as a function of the single variable x. The time-independent, one-dimensional Schrödinger equation is explicitly written as

$$H\psi(x) = E\psi(x). \tag{5.3}$$

We now fill out the expression for the Hamiltonian, H. For a system in which mechanical energy is conserved, H is simply the sum of the kinetic and potential energies. The kinetic energy term, T, in one dimension is $(-\hbar^2/2m)(d^2/dx^2)$. We write the potential energy, U, to explicitly depend on position, and so be a function of x:

$$H\psi(x) = [T + U(x)]\psi(x) = -\frac{\hbar^2}{2m}\frac{d^2\psi(x)}{dx^2} + U(x)\psi(x) = E\psi(x). \tag{5.4}$$

We can collect up the terms in $\psi(x)$ on one side of the equation and then write

$$\frac{d^2\psi(x)}{dx^2} = \frac{2m}{\hbar^2}[U(x) - E]\psi(x). \tag{5.5}$$

Let's spend a moment examining this equation. It is an equation about the function ψ, which is a function of the single variable x. The equation is a differential equation, involving differentiation of ψ by x; more precisely, it is a second-order differential equation, involving differentiation of ψ twice by x: that sums up the left-hand side, $d^2\psi(x)/dx^2$. On the right-hand side appear the mass of the particle, m, and the quantum constant, the reduced Planck constant, $\hbar$. The term in square brackets involves two energies: the potential energy, which is a function of the position, $U(x)$, and the constant energy E. The energy difference, $U(x) - E$, is what is important, rather than $U(x)$ and E separately. Finally, the wavefunction itself, $\psi(x)$, appears.

The key to understanding quantum physics problems is to solve the Schrödinger equation – Equation (5.5), if the problem is time-independent and in one dimension. Solving the equation yields both the wavefunction for all values of x and the constant E. Often there are a number of solutions, characterised by separate values of E, and a different wavefunction for each. We enumerate these as E_n and $\psi_n(x)$. Our task now will be to apply the Schrödinger equation to systems that exhibit terahertz phenomena.

5.3 Free particle

Before turning to physical systems of practical interest, let us set a foundation by briefly considering a very simple physical system, a **free particle**. To start right at the beginning, by definition a particle has no dimensions; we might emphasise this by describing it as a point particle. The particle is characterised by its mass, m. (Throughout I am assuming that the speeds involved are small compared to the speed of light, so relativistic effects may be ignored.)

From the point of view of classical physics, to say a particle is free is to say it is in a region of constant potential U; equivalently, that it is subject to no force, F; equivalently, that there is no acceleration, a, as follows from Newton's second law; equivalently, that the velocity, v, is not changing; equivalently, neither the speed, s, nor the direction of the motion is changing. In other words – the words of Newton's first law – the particle continues at rest or at a uniform motion along a straight line. Since the motion is along a single, straight line, by definition the motion of a free particle is one-dimensional. The momentum, p, is given by $p = mv$ and the energy, E, is given by $E = \frac{1}{2}mv^2$. Neither change.

From the point of view of quantum physics, the momemtum is given by the de Broglie relation:

$$p = \frac{h}{\lambda}. \tag{5.6}$$

Implicit in this formulation is the identification of the particle with a wave, of wavelength λ. The energy is given by

$$E = \frac{p^2}{2m} = \frac{h^2}{2m\lambda^2} = \frac{\hbar^2\sigma^2}{2m} = \frac{\hbar^2 k^2}{2m} = hf = \hbar\omega. \quad (5.7)$$

This is illustrated in Figure 5.1.

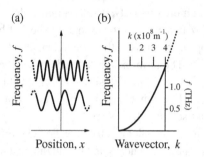

Figure 5.1 A free particle. (a) A free particle may be at any position (horizontal axis) and have any frequency (vertical axis). Higher frequencies correspond to shorter wavelengths. (b) The dispersion curve for a free particle. *Dispersion* relates to how frequency depends on angular wavenumber; if the frequency varies directly with angular wavenumber ($f \propto k$), then there is no dispersion. For a free particle, the frequency varies with the square of the angular wavenumber ($f \propto k^2$). In other words, the dispersion curve is a parabola. The constant of proportionality is $\hbar/4\pi m$, where m is the mass of the particle. Numerically, f [in terahertz] = 0.0921 (k [in units of 10^8 m^{-1}])^2 for the free electron. A frequency of 1 THz corresponds to an angular wavenumber of about 3.3×10^8 m^{-1}. Numerical values for the free electron are given on the top and right axes.

Example 5.1 Free electron, 1 THz

Determine the energy, momentum, angular wavenumber, wavelength and speed of a free electron characterised by a frequency of 1 THz.

Solution

We first calculate $E = hf = 6.63 \times 10^{-22}$ J. Then $p = \sqrt{2mE} = 3.47 \times 10^{-26}$ kg-m/s, $k = p/\hbar = 3.29 \times 10^8$ /m, $\lambda = 2\pi/k = 19.1$ nm and $v = p/m = 38.1$ km/s.

Exercise 5.1 Free electron, 0.1 THz

Determine the energy, momentum, angular wavenumber, wavelength and speed of a free electron characterised by a frequency of 0.1 THz. ∎

Exercise 5.2 Free electron, 10 THz

Determine the energy, momentum, angular wavenumber, wavelength and speed of a free electron characterised by a frequency of 10 THz. ∎

Exercise 5.3 Free proton, 0.1 THz

Determine the energy, momentum, angular wavenumber, wavelength and speed of a free proton characterised by a frequency of 0.1 THz. Which properties does it share with a free electron characterised by the same frequency? ∎

Let us now see how the one-dimensional, time-independent Schrödinger equation (Equation 5.5) applies to the free particle. For a free particle, there is no confining potential. In other words, $U(x)$ is zero for all values of x. So Equation (5.5) becomes

$$\frac{d^2\psi(x)}{dx^2} = \frac{2m}{\hbar^2}[0-E]\psi(x) = -\frac{2mE}{\hbar^2}\psi(x). \tag{5.8}$$

The equation to solve is of the form

$$\frac{d^2y}{dx^2} = -Cy. \tag{5.9}$$

Here I have replaced $\psi(x)$ with y and $2mE/\hbar^2$ with C, a positive constant, to simplify the equation and bring out its essential structure. We need to find a function, y, which we differentiate twice with respect to x, and end up back with the original function (except that the sign has changed and it is now multiplied by a factor C). What function, when differentiated twice, gives back itself? The answer is the sine function (or the cosine function; or a combination of these). So let's try for the solution of Equation (5.8):

$$\psi(x) = A\sin(kx+\delta). \tag{5.10}$$

Here three constants appear: the amplitude A, the angular wavenumber k and the initial phase factor δ. The values of these constants will be determined as we continue to solve the equation.

Differentiating once and twice with respect to x yields

$$\frac{d\psi(x)}{dx} = kA\cos(kx+\delta), \tag{5.11}$$

$$\frac{d^2\psi(x)}{dx^2} = -k^2 A\sin(kx+\delta) = -k^2\psi(x). \tag{5.12}$$

So we have confirmed that the proposed solution Equation (5.10) to Equation (5.8) does the job, provided the identification is made that $k^2 = 2mE/\hbar^2$. This condition on k is simply another way of arriving at the prescriptions of Equation (5.6) and Equation (5.7). The wavefunction for the free particle is then

$$\psi(x) = A\sin\left(\sqrt{\frac{2mE}{\hbar^2}}x + \delta\right). \tag{5.13}$$

Let us look at the essential features of this wavefunction. The dependence of ψ on x is given by the sine function. If we prefer, we could use a cosine form, by writing $\psi(x) = A\cos(kx+\delta')$, where $\delta' = \delta - \pi/2$ radians; or avoid the initial phase factor altogether, by writing $\psi(x) = B\sin(kx) + C\cos(kx)$; all the expressions are equivalent. In general, in solving a second-order differential equation, two constants arise. In this case, the two constants are A and δ. The constants are usually determined using the boundary conditions of the problem. The initial phase factor, δ, relates to where on the cycle of the sine function we are for a given value of x. For convenience, we can choose the origin of the x axis to coincide with a position where δ is zero; then ψ is zero, since $\sin(0) = 0$. The amplitude, A, relates to the size of the wavefunction. This is a little problematic when we are dealing with infinite space. Let us first consider that the particle is restricted to a region of length L, which, for convenience, we will presume corresponds to an integral number of wavelengths ($L = n\lambda = 2\pi n/k$) starting at initial phase factor $\delta = 0$, and then allow this L to approach infinity. The normalisation condition is

$$\int_{x=0}^{L} \psi(x)^* \psi(x) dx = \int_{x=0}^{L} A \sin(kx) A \sin(kx) dx = 1. \tag{5.14}$$

We may write the integral as

$$\int_{x=0}^{n\lambda} A^2 \sin^2(kx) dx = A^2 \int_{x=0}^{2\pi n/k} \sin^2(kx) dx. \tag{5.15}$$

The integral is of a standard form and has the solution

$$\int_{x=0}^{2\pi n/a} \sin^2(ax) dx = \frac{x}{2} - \frac{\sin(2ax)}{4a} \Big|_0^{2\pi n/a} = \left(\frac{2\pi n/a}{2} - 0\right) - (0 - 0) = \pi n/a. \tag{5.16}$$

Since here $\pi n/a = \pi n/k = L/2$, the normalisation condition becomes

$$A = \sqrt{\frac{2}{L}}. \tag{5.17}$$

As expected, as L increases, A decreases (but we run into conceptual difficulties if we try to take L all the way to infinity and so make A vanish).

Putting together these considerations for the values of the constants A and δ, we may write the wavefunction for the free particle as

$$\psi(x) = \sqrt{\frac{2}{L}} \sin\left(\sqrt{\frac{2mE}{\hbar^2}} x\right) = \sqrt{\frac{2}{L}} \sin(kx). \tag{5.18}$$

Speaking of conceptual difficulties, the 'free particle' is something of an oxymoron from the quantum perspective. The momentum is precisely defined (Equation 5.6), but the position is indeterminate. We can say exactly what the wave property – the wavelength – is, but cannot determine the position at all; it exists everywhere in space. Some particle.

5.4 Infinitely high, square potential well

As the next most simple quantum system, let us consider a particle that is free to move in one region of space, but forbidden to move anywhere else. The **square well**, as shown in Figure 5.2, is a good approximation to many real, physical systems.

5.4 Infinitely high, square potential well

We will start with a single dimension, and call that x. We will say one end of the allowed motion is at $x = 0$ and the other end of the allowed motion is at $x = L$.

In terms of potential U, there is no change of potential in the range $x = 0$ to $x = L$. In other words, there is no force acting on the particle in this region. In other words, within that region, the particle can move freely. At either end of this region, there is a potential barrier. It is abrupt. It goes straight up. It is impossibly high. There is no way the particle can climb the barrier and escape. It is like the vertical wall of an infinitely deep well. The particle is trapped in this well, cannot escape, but can move freely within its prison.

We write down the potential for the three regions as

$$U(x) \to \infty \quad x < 0, \tag{5.19}$$

$$U(x) = 0 \quad 0 < x < L, \tag{5.20}$$

$$U(x) \to \infty \quad x > L. \tag{5.21}$$

From the point of view of wave mechanics, the stationary states that correspond to the solution of the Schrödinger equation relate to standing waves in the well. The wavefunction must vanish outside the well, so the edges of the well ($x = 0$ and $x = L$) must be nodes.

If the possible states are enumerated in terms of total nodes, we can have either two nodes (one at each end of the well), three nodes (including one in the middle of the well), four nodes, five nodes, and so on. These correspond to half, one, one and a half, two, two and a half, and so on, wavelengths in the well; we may list these as $\lambda_1/2 = L$, $\lambda_2 = L$, $3\lambda_3/2 = L$, $2\lambda_4 = L$, $5\lambda_5/2 = L$, or

$$\lambda_n = \frac{2L}{n}, \tag{5.22}$$

where the label n takes on the values $1, 2, 3, \ldots$. From this equation, $\lambda_1 = 2L$, so we may write any of the wavelengths in terms of the first:

$$\lambda_n = \frac{1}{n}\lambda_1. \tag{5.23}$$

The series progresses inversely with the index n.

Immediately, we see the system is quantised. Only certain values of λ are allowed. Others are forbidden. As we shall see, this implies only certain angular wavenumbers are allowed, and only certain energies are allowed. The allowed states are enumerated using the quantum number n.

We may express this quantisation condition in terms of angular wavenumber rather than wavelength through the relation $k = 2\pi/\lambda$:

$$k_n = \frac{2\pi}{\lambda_n} = n\frac{\pi}{L}. \tag{5.24}$$

A most convenient property of this equation is that the allowed values of k progress exactly as the allowed values of n,

$$k_n = nk_1. \tag{5.25}$$

There is a fundamental value of $k_1 = \pi/L$. The other allowed values are 2, 3, 4, ... times this fundamental value.

Likewise, the momentum is quantised according to

$$p_n = \hbar k_n = n\frac{h}{2L} = np_1. \tag{5.26}$$

Next, let's write down the allowed values of the energy of the particle in the well from $E = \hbar^2 k^2/2m$:

$$E_n = \frac{\hbar^2 k_n^2}{2m} = n^2\frac{h^2}{8mL^2}. \tag{5.27}$$

The first thing to note about this equation is that energy is quantised: the particle in the well can have certain energies and no others. The second thing to note is that the energy increases quadratically with the state quantum number. The fundamental energy is $E_1 = h^2/8mL^2$. Successive states have energies 4, 9, 16, 25, ... times the fundamental energy:

$$E_n = n^2 E_1. \tag{5.28}$$

The difference in energy between the states characterised by indices m and n is therefore

$$\Delta E_{n \to m} = E_m - E_n = (m^2 - n^2)E_1 = (m^2 - n^2)\frac{h^2}{8mL^2}. \tag{5.29}$$

Given these energies, it is now a simple matter of writing down the frequency of photons with equivalent energy through the relation $E = hf$. Dividing the expressions for E by h yields

$$f_n = n^2\frac{h}{8mL^2} = n^2 f_1. \tag{5.30}$$

For numerical calculations, it is convenient to express the frequency in terms of terahertz, masses in terms of electron masses and lengths in terms of nanometres. In these units the equation reads

$$f_n \text{ [in terahertz]} = n^2 \frac{90.92}{m \text{ [in electron masses]}(L \text{ [in nanometres]})^2}. \tag{5.31}$$

Likewise, for frequency differences,

$$\Delta f_{n \to m} = (m^2 - n^2)f_1 = (m^2 - n^2)\frac{h}{8mL^2}, \tag{5.32}$$

and for numerical calculations,

$$\Delta f_{n \to m} \text{ [in terahertz]} = (m^2 - n^2)\frac{90.92}{m \text{ [in electron masses]}(L \text{ [in nanometres]})^2}. \tag{5.33}$$

5.4 Infinitely high, square potential well

Example 5.2 Confined electron

What is the characteristic fundamental frequency of an electron confined to a length of the order of the size of the atom, 0.1 nm?

Solution

We use Equation (5.31) with m, in electron masses, being 1, and L, in nanometres, being 0.1. Then f_1 is 9092 THz.

Inside the well, the particle moves as a free particle, so the wavefunction is given by Equation (5.18). The particle cannot get out of the well; the probability of finding it outside the well is zero, and so the wavefunction outside the well is zero. Thus

$$\psi(x) = 0 \quad x < 0, \tag{5.34}$$

$$\psi_n(x) = \sqrt{\frac{2}{L}} \sin(k_n x) = \sqrt{\frac{2}{L}} \sin\left(\frac{n\pi}{L}x\right) \quad 0 < x < L, \tag{5.35}$$

$$\psi(x) = 0 \quad x > L. \tag{5.36}$$

The importance of the infinite well to terahertz physics is that structures can be fabricated in which the energy spacings are in the range of terahertz-frequency photons. The example and the exercises illustrate this. Typically, such structures are made by sandwiching a few atomic layers of one semiconductor, which serves as the bottom of the well, between another semiconductor, which serves as the sides of the well. In one dimension, normal to the layers, electrons are confined. In the other two dimensions, within the layer, the electrons can move freely. This idea can be continued to confinement in two dimensions, in which case the structure is called a **quantum wire**. Electrons can move freely along the length of the quantum wire, but cannot escape from the sides. If the confinement is extended to the third dimension as well, the entity is known as a **quantum dot**.

Example 5.3 The quantum well (QW)

An electron is confined to a length of 10 nm. Give the first three allowed wavelengths, angular wavenumbers and energies of the electron. Find the frequencies of photons with the same energies.

Solution

We use Equation (5.22) to calculate the wavelengths. They are $\lambda_1 = 20$ nm, $\lambda_2 = 10$ nm, $\lambda_3 = 6.67$ nm. Equation (5.24) yields the angular wavenumbers: $k_1 = 3.14 \times 10^8$ m^{-1}, $k_2 = 6.28 \times 10^8$ m^{-1}, $k_3 = 9.42 \times 10^8$ m^{-1}. Finally, Equation (5.27) gives the energies: $E_1 = 6.02 \times 10^{-22}$ J, $E_2 = 24.1 \times 10^{-22}$ J, $E_3 = 54.2 \times 10^{-22}$ J, corresponding to photon frequencies of 0.9, 3.6 and 8.2 THz.

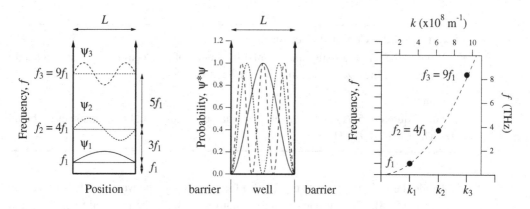

Figure 5.2 An infinite one-dimensional square well – an ideal quantum well. The well is of length L. It is bounded by infinitely high potential barriers. Particles can be found inside the well, but not outside. The energy states, the angular wavenumbers and the frequencies are all quantised. The dispersion relationship is the same as for a free particle, namely, parabolic; but the particle can only have discrete values of the angular wavenumber. The right and top axes give numerical values for an electron confined to a well of width $L = 10$ nm.

Exercise 5.4 Effective mass

An electron in the semiconductor GaAs has an effective mass about 7% of the usual electron mass. Consider an electron confined to a 1 nm length of GaAs. What are the first three energy levels? What is the corresponding photon energy corresponding to a transition between the first and second energy states? ∎

Exercise 5.5 Quantum wire

A particle is confined in the x direction to a length of L_x and in the y direction to a length L_y. Show that the allowed energies are characterised by two quantum numbers, n_x and n_y, according to

$$E(n_x, n_y) = n_x^2 \frac{h^2}{8mL_x^2} + n_y^2 \frac{h^2}{8mL_y^2}. \tag{5.37}$$

∎

Exercise 5.6 Quantum cube or quantum dot

Show that a particle of mass m, confined in three perpendicular directions to a length L, has energy states given by

$$E(n_x, n_y, n_z) = (n_x^2 + n_y^2 + n_z^2) \frac{h^2}{8mL^2}. \tag{5.38}$$

Determine the energies of the (1,1,1), (1,1,2), (1,2,2) and (2,2,2) states. ∎

5.5 Finite, square potential well

The infinite potential well is a good approximation for many physical systems. In practice, the potential well will not be infinite, although it may be very large. An example is shown in Figure 5.3.

5.5 Finite, square potential well

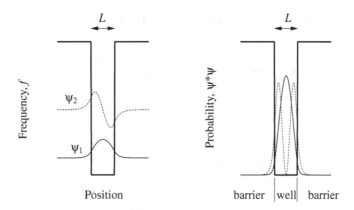

Figure 5.3 A finite one-dimensional square well – a real quantum well. The well is of length L. It is bounded by finitely high potential barriers. Inside the well, the energy states, the angular wavenumbers and the frequencies are all quantised. The particle may have sufficient energy to escape the well. The figure approximates the energy states of an electron, of effective mass 0.067, in a quantum well of width $L = 10$ nm made of GaAs sandwiched between barriers made of $Al_{0.3}Ga_{0.7}As$. The first energy state is at 32 meV, corresponding to about 8 THz, and the second energy state is at 122 meV, corresponding to about 30 THz. These are the only two states confined in the well.

The Schrödinger equation of the finite potential well can be solved by looking at the separate regions in space, then bringing the solutions together so that they join up. In more detail: the solution within the well has the form of the free-particle solution, and can be written in terms of a combination of sines and cosines; outside the well the solution has the form of a decaying probability, and can be written in terms of an exponential function. There is a chance the particle will leak or *tunnel* outside the well.

The problem can be solved by analytic or numerical methods and the details of the solution depend on the initial parameters. The energies are still quantised, however, and still have the same form as before:

$$E(v_n) = v_n^2 \frac{h^2}{8mL^2}. \tag{5.39}$$

The important thing is that, although the labels n proceed as before, counting up from 1, the values of v_n are not necessarily integers. The exact values of v_n depend on the parameters of the problem (L, m and the height of the potential well, U_0).

In the case of the infinite potential well, there are an infinite number of energy states. In the case of the finite well, there are only a finite number of energy states. If the particle in the well has sufficient energy it will escape the well. In the extreme case of a very deep and narrow well, there is only one energy level within the well. If the particle is found in such a well, it must have this energy.

5.6 Multiple, finite, square potential wells

What happens when we have two square potential wells side-by-side? We will assume the wells are of similar physical size and not too distant. Exactly what happens depends on the depth (in terms of energy) of the wells and their width and separation from each other (in the x direction), but the main feature of apposition is that each separate energy level now splits into two. So, where in a single well there were energies E_1, E_2, E_3, and so on, in the double well there are energies E_{1a}, E_{1b}, E_{2a}, E_{2b}, E_{3a}, E_{3b}, and so on. The lower energy pairs (for example, E_{1a}, E_{1b}) are more closely spaced than the higher pairs (for example, E_{3a}, E_{3b}).

As we increase the number of wells, the number of closely linked energy levels increases. For example, in a group of six identical potential wells, there will be six energy levels close to the energy level E_1 of an isolated well, and six energy levels close to the original energy level E_2.

5.6.1 Energy bands

We can continue this process for a larger and larger number of potential wells. In the extreme, we can consider an infinite number of potential wells. An infinite number of identical potential wells is a good model for the electronic structure of a crystal lattice. Each atom has associated with it an attractive potential for an electron. Depending on the depth of the wells and the spacing of the atoms, a large number of energy states can cluster together near the original energy states of an isolated atom. This large number of energy states in close proximity is called an *energy band*.

5.6.2 Semiconductors, superconductors, superlattices

It may be that there is a significant gap in energy between one energy band and the next energy band. This gap in energy is referred to as a *bandgap*.

In many semiconductors – Si, Ge, GaAs, for example – there is a bandgap between the states that the electrons occupy and the next available energy band. The bandgap is of the order of 1 eV, or about 250 THz.

The bandgap of a semiconductor is its most important characteristic. The energy associated with the bandgap is normally much larger than the energies associated with the terahertz range. Even in so-called *narrow gap* semiconductors, where the bandgap is of the order of 0.1 eV, the corresponding photon frequency is about 25 THz, so beyond what we normally consider the terahertz range. There is some interest in reducing the bandgaps of semiconductors, for example, by doping GaAs with Bi or, on the other hand, slightly opening the gap in the gapless material graphene. Although the emission or detection of visible light utilising the bandgap in a variety of semiconductors is a well-established technique, the direct use of bandgaps for terahertz emission or detection is limited.

A somewhat related bandgap is found in the energy spectrum of superconductors. For example, classic superconductors have critical temperatures around a few kelvin. The bandgap is typically about four times this thermal energy. So superconductors are studied by terahertz radiation, and some schemes have been devised to generate terahertz or sub-terahertz radiation from superconductors.

To recapitulate: placing potential wells together gives *bands* of allowed energies where previously only a single energy state existed; a large number of potential wells, as in a crystal lattice, will produce energy bands. The natural crystal lattices that are so well described in this way have been joined in the last decades by artificially grown, larger-scale lattices, which still show the same quantum effects. These are called *superlattices*.

A superlattice is typically grown with ten atomic layers of one crystal structure followed by ten layers of another crystal structure, and so on, over and over again. While the confinement of an electron to the size of about an atom (0.1 nm) leads to states quantised in steps of about an eV, the looser confinement to, say, ten atomic layers will lead to a 10^2 reduction in energy, to the 10 meV, or terahertz range. Superlattices are therefore important structures for terahertz emission and detection.

5.7 Oscillations of molecules

We might consider a single atom to be a hard sphere. (Of course, there is more to the atom than this, and if we look inside we will see electrons, and if we look closer still we will see nucleons. Oscillations involving electrons will be dealt with elsewhere in this book and oscillations involving neutrons won't be dealt with at all. For the moment, we will consider the atom to have no internal features.) There is not much scope for it to oscillate. It could bounce up and down in space, but let's assume it stays in the same place. It could rotate on its axis, but we will ignore this (more precisely, the angular momentum involved is small).

Let's move up in the world and consider two atoms joined together. We can imagine the atoms joined by a stick – then we have the ball-and-stick model. We'll imagine the stick can flex. If you prefer, you can imagine the atoms to be joined by a spring.

Two atoms joined by a spring have many more ways to oscillate than a single atom. Although the atoms could rotate with the direction of the spring, z, being the rotation axis, the angular momentum involved is small, and we will ignore this (as we did for the single atom). The atoms can rotate in a direction perpendicular to this (say, the x direction) and the angular momentum will be large. We may define a third direction, y, perpendicular to the other two. A general rotation along an axis lying in the x-y plane can then be defined in a straightforward way. It turns out that the rotational frequencies of many small molecules occur in the terahertz range.

There is a another simple way in which two atoms joined by a spring may oscillate. The spring may expand and contract. The atoms move closer and farther apart. This lengthening and compressing of the spring is called a *stretching* oscillation.

The situation becomes more involved when there are three atoms in the molecule. For example, the whole molecule might begin in a straight line, and the two end atoms both move off this central line in the same direction. This is called a *bending* oscillation.

5.7.1 Rotational motion

Rotational motion is quantised. This means that rotation is not permitted at any frequency, but only at certain frequencies.

The concept of quantisation is foreign to everyday notions of rotation. In the world of our experience, there is nothing to prevent a wheel turning at one frequency or another or anywhere in between. For example, as a car starts from rest, the wheels are not turning at all, then start to turn slowly, then turn faster and faster and so on. The change in frequency appears to be continuous.

Things are different in the quantum world. An object may be stationary, or rotate at a minimum frequency, or at other specified frequencies, but nowhere in between.

Analogies to try to explain the microscopic quantum world in terms of the macroscopic everyday world always have shortcomings. But here is one. Vinyl records were made to operate at certain fixed rotation speeds: 78 revolutions per minute, 45 rpm and $33\frac{1}{3}$ rpm. So record players only have discrete frequencies ($33\frac{1}{3}$, 45 and 78 rpm) at which they operate. Here is another analogy. Imagine you are pedalling a bicycle along a flat road at a constant speed of, say, 20 km/h. You then change gears, but maintain the same speed of 20 km/h. You will be pedalling at a different frequency. The frequency is not arbitrary, but is set by the ratio of the gears. So (all the while maintaining the same road speed), only a discrete set of pedalling frequencies is permitted, and these are determined by the mechanics of the bicycle.

Rotational motion is most often encountered in molecules isolated from each other, as in a gas. In liquids the rotational motion is rapidly damped out in collisions with nearby molecules. In solids, the atoms bind together, and isolated molecules are not found.

In describing rotation, it is useful to introduce the concept of moment of inertia, I. The moment of inertia can be understood to play the role in rotational motion that mass plays in translational motion. Just as the product of the mass and the acceleration is equal to the force in translational motion, the product of moment of inertia and angular acceleration is equal to the angular force, or torque, in rotational motion. The moment of inertia for a particle of mass m rotating about a point at a distance r away is

$$I = mr^2. \tag{5.40}$$

As we shall see in a moment, a natural energy unit that arises in the quantum theory of rotation is

$$\frac{\hbar^2}{2I}, \tag{5.41}$$

where $\hbar$ is the Planck constant divided by 2π. Let us make a rough estimate of this quantity for a simple molecule. The simplest molecule I can think of is one simplest atom, hydrogen, joined to another, H_2. Let us assume, for simplicity, that the separation of the two atoms is one tenth of a nanometre, and take the mass of each hydrogen atom to be 1.7×10^{-27} kg. We assume the two atoms are each rotating about the common centre of mass, located halfway between them. Then the moment of inertia of the molecule is $2 \times 1.7 \times 10^{-27}$ kg $\times (0.1/2 \times 10^{-9}$ m$)^2 = 9 \times 10^{-48}$ kg-m^2. Since $\hbar^2 = 1.11 \times 10^{-68}$ kg^2 m^4 s^{-2}, the energy term $\hbar^2/2I$ amounts to 6.6×10^{-22} J, or 4 meV, or to the energy of

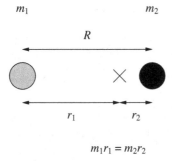

Figure 5.4 A system of two particles. The system is characterised by three parameters: the mass of the first particle, m_1, the mass of the second particle, m_2 and the separation of the two particles, R (as shown at the top of the figure). A point between the two particles may be identified as the *centre of mass* (as indicated by the cross). The distance from the first particle to the centre of mass is r_1 and the distance to the second particle from the centre of mass is r_2. These distances are in inverse ratio to the masses, $r_1 : r_2 = m_2 : m_1$, or $m_1 r_1 = m_2 r_2$. The distance from the first particle to the centre of mass plus the distance of the second particle to the centre of mass is the total distance between the two particles, so $r_1 + r_2 = R$. The total mass of the system, M, is $M = m_1 + m_2$. The *reduced mass* of the system, μ, is defined by the equation $\mu M = m_1 m_2$. That is, the product of the reduced mass and the total mass is equal to the product of the two masses. Thus $\mu = m_1 m_2 / M = m_1 m_2 / (m_1 + m_2)$.

a photon with frequency of 1 THz. This is the lightest molecule. For heavier molecules, with larger moments of inertia, the corresponding energy will be less.

Exercise 5.7 Reduced mass

In dealing with diatomic molecules it is convenient to use the concept of *reduced mass*. See Figure 5.4. Consider two particles, of masses m_1 and m_2. (The total mass of the two-particle system is $M = m_1 + m_2$.) The two particles are separated by a distance R. We define the centre of mass to lie on the line between the two particles. The distance from the centre of mass to the first particle is r_1 and the distance from the centre of mass to the second particle is r_2. From the definitions of R, r_1 and r_2,

$$R = r_1 + r_2. \tag{5.42}$$

The distances r_1 and r_2 are defined to be in inverse proportion to the corresponding masses, or

$$m_1 r_1 = m_2 r_2. \tag{5.43}$$

(a) Eliminate r_2 from Equations (5.42) and (5.43) to obtain an expression for r_1,

$$r_1 = \frac{m_2}{m_1 + m_2} R. \tag{5.44}$$

Obtain the cognate expression for r_2,

$$r_2 = \frac{m_1}{m_1 + m_2} R. \tag{5.45}$$

(b) The moment of inertia for a particle of mass m rotating around an axis at distance r is

$$I = mr^2. \tag{5.46}$$

Thus the moment of inertia for particle 1 rotating around the centre of mass is

$$I_1 = m_1 r_1^2. \tag{5.47}$$

Similarly, the moment of inertia for particle 2 rotating around the centre of mass is

$$I_2 = m_2 r_2^2. \tag{5.48}$$

The total moment of inertia for the two-particle system is the sum of the individual moments of inertia,

$$I_t = I_1 + I_2 = m_1 r_1^2 + m_2 r_2^2. \tag{5.49}$$

Now use Equations (5.44) and (5.45) to eliminate r_1 and r_2 from Equation (5.49) to obtain

$$I_t = \mu R^2, \tag{5.50}$$

where

$$\mu = \frac{m_1 m_2}{m_1 + m_2} \tag{5.51}$$

is the *reduced mass* of the system of two particles. Points to note:

- Equation (5.50) has the same form as Equation (5.46). The mass of a single particle (m) has been replaced by the reduced mass of the two particles (μ) and the distance of the single particle from the axis of rotation (r) has been replaced by the separation of the two particles (R).
- The reduced mass of two particles (Equation 5.51) is the product of the masses divided by the sum of the masses.
- The reduced mass of two particles (Equation 5.51) is a mass and has SI units of kg.
- The reduced mass of two particles is less than the mass of either particle. Writing $\mu = m_1 \times \frac{m_2}{m_1 + m_2}$ we see that $\mu < m_1$, since $\frac{m_2}{m_1 + m_2} < 1$. Likewise, $\mu < m_2$.

(c) Special cases.
(i) Equal masses, $m_1 = m_2 = m$.
Show that, in the case of equal masses,

$$\mu = \frac{m_1}{2} = \frac{m_2}{2} \tag{5.52}$$

and

$$I_t = \frac{m_1}{2} R^2 = \frac{m_2}{2} R^2. \tag{5.53}$$

(The last result may be obtained directly from Equation (5.49) by substituting $r_1 = r_2 = R/2$.)

(ii) Very unequal masses, $m_1 \ll m_2$.
Show that, for $m_1 \ll m_2$,

$$\mu \sim m_1 \tag{5.54}$$

and

$$I_t \sim m_1 R^2. \tag{5.55}$$

(In this case, the more massive particle hardly moves, and acts as a centre about which the less massive particle revolves. The massive particle may as well not be there.) ∎

For an object of an arbitrary shape, three moments of inertia, I_A, I_B and I_C, can be determined for rotation about the centre of mass around three perpendicular axes. The order A, B and C corresponds to non-decreasing moments of inertia. We will now divide the possibilities up into several subsets depending on the relative sizes of I_A, I_B and I_C. In each case I will write down the allowed energies. These arise from the solution of the Schrödinger equation in each case. I will not give the derivations, only the results.

Spherical rotor: $I_A = I_B = I_C = I$
This is the most symmetric case imaginable. The three moments of inertia, one corresponding to each of three perpendicular directions, are the same. A spherically symmetric object will have moments of inertia like this. In this case, the x, y and z directions may be chosen respectively to lie along the A, B and C directions, and the Hamiltonian may be written

$$H = \frac{1}{2}\left[\frac{p_x^2}{I_A} + \frac{p_y^2}{I_B} + \frac{p_z^2}{I_C}\right]. \tag{5.56}$$

Rotational states are characterised by a quantum number J that takes on the integer values

$$J = 0, 1, 2, 3, \ldots. \tag{5.57}$$

The energy of the rotational state labelled J is

$$E(J) = \frac{\hbar^2}{2I}J(J+1). \tag{5.58}$$

The selection rule for transitions between states is

$$\Delta J = 1. \tag{5.59}$$

Symmetric rotor, oblate: $I_A = I_B < I_C$
Symmetric rotor, prolate: $I_A < I_B = I_C$
In a symmetric rotor, the moments of inertia are the same in two directions, but different in the third. The symmetric rotor may be further divided into the oblate and prolate versions. The oblate symmetric rotor has the same moment of inertia in two directions, such as two perpendicular directions in the plane of a plate, and a larger moment of inertia in the third direction, namely perpendicular to the plate. The prolate symmetric

rotor has a smaller moment of inertia in the third direction, such as along the long axis of a rugby ball. For the prolate symmetric rotor, the Hamiltonian may be written

$$H = \frac{1}{2}\left[\frac{p^2}{I_B} + p_z^2\left(\frac{1}{I_A} - \frac{1}{I_B}\right)\right]. \tag{5.60}$$

Rotational states are now characterised by the two quantum numbers

$$J = 0, 1, 2, 3, \ldots. \tag{5.61}$$

and

$$K = -J, -J+1, -J+2, \cdots -2, -1, 0, 1, 2, \ldots, J-2, J-1, J. \tag{5.62}$$

The energy of the rotational state labelled J, K is

$$E(J, K) = \frac{\hbar^2}{2I_B}J(J+1) + \frac{\hbar^2}{2}\left(\frac{1}{I_A} - \frac{1}{I_B}\right)K^2. \tag{5.63}$$

The selection rules for transitions between states are

$$\Delta J = 1 \tag{5.64}$$

and

$$\Delta K = 0. \tag{5.65}$$

Linear rotor: $I_A = 0 < I_B = I_C = I$

The moments of inertia are the same in two directions, but negligible in the third. Rotational states are labelled with the single quantum number

$$J = 0, 1, 2, 3, \ldots, \tag{5.66}$$

and the energies of the states

$$E(J) = \frac{\hbar^2}{2I}J(J+1), \tag{5.67}$$

as for the spherical case. Again, the selection rules for transitions between states are

$$\Delta J = 1. \tag{5.68}$$

Asymmetric rotor: $I_A \neq I_B \neq I_C$

This case is complicated and will not be considered here.

Transitions between states

For each of the rotors considered – spherical, symmetric or linear – the selection rule for transition between states of different values of J is that $\Delta J = \pm 1$. Let us consider a transition in which the rotor increases in energy. In the transition from the initial state J to the final state $J + 1$ the change in energy is

$$\Delta E = E(J+1) - E(J) = \frac{\hbar^2}{2I}[J+1]([J+1]+1) - \frac{\hbar^2}{2I}J(J+1) = 2\frac{\hbar^2}{2I}(J+1). \quad (5.69)$$

(In the case of the symmetric top there is also a term involving the quantum number K, but this disappears when the energy difference is calculated.) So the transition energy is directly proportional to the label of the final state.

The energies of the states can be listed explicitly for the various quantum numbers J:

$$J = 0 \qquad E(J) = 0$$
$$J = 1 \qquad E(J) = 2\tfrac{\hbar^2}{2I}$$
$$J = 2 \qquad E(J) = 6\tfrac{\hbar^2}{2I}$$
$$J = 3 \qquad E(J) = 12\tfrac{\hbar^2}{2I}$$
$$\vdots$$

Likewise, the energy differences for transitions between the states can be listed explicitly:

$$J = 0 \to J = 1 \qquad \Delta E = 2\tfrac{\hbar^2}{2I}$$
$$J = 1 \to J = 2 \qquad \Delta E = 4\tfrac{\hbar^2}{2I}$$
$$J = 2 \to J = 3 \qquad \Delta E = 6\tfrac{\hbar^2}{2I}$$
$$J = 3 \to J = 4 \qquad \Delta E = 8\tfrac{\hbar^2}{2I}$$
$$\vdots$$

Example 5.4   HCl

Estimate the energies of the first six rotational states of HCl and the transition energies between them.

Solution

We will take as input data for the masses of the H and Cl ions and the separation between them

$$m_H = 1.67 \times 10^{-27} \text{ kg},$$
$$m_{Cl} = 58.9 \times 10^{-27} \text{ kg},$$
$$R = 0.127 \text{ nm}.$$

Using Equation (5.51) to calculate the reduced mass of the molecule, we find $\mu = 1.627 \times$

10^{-27} kg. (Note, this is rather close to m_H, since $m_{Cl} \gg m_H$.) We now use Equation (5.50) to find the moment of inertia to be $I = 2.62 \times 10^{-47}$ kg m². We next calculate the term $\hbar^2/2I = 2.12 \times 10^{-22}$ J, corresponding to 1.32 meV, the energy of photons of frequency 0.319 THz.

$J = 0$	$E(0) = 0$ THz		
$J = 1$	$E(1) = 0.64$ THz	$J = 0 \to 1$	$\Delta E(0 \to 1) = 0.64$ THz
$J = 2$	$E(2) = 1.92$ THz	$J = 1 \to 2$	$\Delta E(1 \to 2) = 1.28$ THz
$J = 3$	$E(3) = 3.83$ THz	$J = 2 \to 3$	$\Delta E(2 \to 3) = 1.92$ THz
$J = 4$	$E(4) = 6.39$ THz	$J = 3 \to 4$	$\Delta E(3 \to 4) = 2.55$ THz
$J = 5$	$E(5) = 9.58$ THz	$J = 4 \to 5$	$\Delta E(4 \to 5) = 3.19$ THz

Exercise 5.8 HF

Estimate the energies of the first six rotational states of HF and the transition energies between them, given the data

$$m_H = 1.67 \times 10^{-27} \text{ kg},$$
$$m_F = 31.5 \times 10^{-27} \text{ kg},$$
$$R = 0.092 \text{ nm}. \blacksquare$$

5.7.2 Vibrational motion

The simplest vibration occurs when two atoms are joined to form a diatomic molecule. Since a straight line may be drawn through any two points, a diatomic molecule is linear. There is one vibrational mode: the two atoms can move toward each other then away from each other along the line that joins them. One can imagine this as the spring joining the atoms alternately stretching and compressing.

We have seen that the rotational motion of a typical diatomic molecule occurs in the terahertz frequency range. The vibrational motion typically occurs at a frequency of one hundred to one thousand times larger. Vibrational transitions are therefore associated more with the infrared and the visible parts of the spectrum rather than the terahertz. Combinations of vibrational and rotational transition do occur commonly in the terahertz range, however, and we will discuss pure vibrational motion as a lead-in to the more involved rotational-vibrational effects.

The energy of the nth level of a quantum harmonic oscillator is given by

$$E(n) = (n + \frac{1}{2})\hbar\omega = (n + \frac{1}{2})hf, \tag{5.70}$$

where the angular frequency of vibration is given in terms of the force constant β and the reduced mass μ as

$$\omega = \sqrt{\frac{\beta}{\mu}}. \tag{5.71}$$

The quantum number n takes on the values

$$n = 0, 1, 2, 3, \ldots, \tag{5.72}$$

and transitions between vibrational states are permitted according to the prescription

$$\Delta n = \pm 1. \tag{5.73}$$

Vibrational motion of polyatomic molecules

Each atom in a molecule has three degrees of freedom, corresponding to motion in each of the three spatial dimensions. A molecule comprising N atoms will thus have $3N$ degrees of freedom. Considering the molecule as a whole, the translation of the molecule (in other words, the motion of centre of mass of the molecule) has three degrees of freedom. Rotation of the molecule as a whole uses another three degrees of freedom. Taking away these degrees of freedom for the molecule as a whole leaves $3N - 3 - 3 = 3N - 6$ degrees of freedom. (For a linear molecule, there is no rotation around the length, so one fewer degrees of freedom to subtract; a linear molecule has $3N - 5$ degrees of freedom.) Thus, a triatomic (nonlinear) molecule has $3 \times 3 - 6 = 3$ degrees of freedom. These are

- symmetric longitudinal stretching,
- antisymmetric longitudinal stretching,
- symmetric transverse stretching (also called *bending*, or *scissoring*).

5.8 Summary

5.8.1 Key terms

free particle, 77
square well, 80

quantum well, 83
quantum wire, 83

quantum dot, 83

5.8.2 Key equations

time-independent Schrödinger equation	$\dfrac{d^2\psi(x)}{dx^2} = \dfrac{2m}{\hbar^2}[U(x) - E]\psi(x)$	(5.5)
free-particle wavefunction	$\psi(x) = \sqrt{\dfrac{2}{L}}\sin(kx)$	(5.18)
energy of particle in a well	$E_n = \dfrac{\hbar^2 k_n^2}{2m} = n^2 \dfrac{h^2}{8mL^2}$	(5.27)
energy of rotational states	$E(J) = \dfrac{\hbar^2}{2I}J(J+1)$	(5.58)
energy of quantum harmonic oscillator	$E(n) = (n + \dfrac{1}{2})\hbar\omega = (n + \dfrac{1}{2})hf$	(5.70)

5.9 Table of symbols, Chapter 5

General mathematical symbols appear in Appendix B. If the unit of a quantity depends on the context, this is denoted '—'.

Symbol	Meaning	Unit
a	general quantity	—
A	general quantity	—
B	general quantity	—
C	general quantity	—
E	energy	J
f	frequency	Hz
h	Planck constant	J Hz^{-1}
$\hbar$	reduced Planck constant	J Hz^{-1}
H	Hamiltonian	J
I	moment of inertia	kg m^2
J	number	[unitless]
k	angular wavenumber	rad m^{-1}
K	number	[unitless]
L	length	m
m	mass	kg
m	integer	[unitless]
n	integer	[unitless]
p	momentum	kg m Hz
r	radius	m
R	radius	m
T	kinetic energy	J
U	potential energy	J
v	speed	m Hz
x	coordinate label	—
y	coordinate label	—
z	coordinate label	—
δ	initial phase	rad
λ	wavelength	m
μ	reduced mass	kg
ν_n	number	[unitless]
ψ	wavefunction	—
ω	angular frequency	rad Hz

6 INTERACTION OF LIGHT AND MATTER

This chapter calls on trigonometry and calculus, both differentiation and integration.

In this chapter, we will see what happens when terahertz-frequency electromagnetic radiation encounters matter.

Imagine driving along a freeway. All the cars are travelling along at the same, high speed, 100 kmh. There is no stopping or turning on the freeway. (It is a little bit boring, really.) Then the freeway ends. The traffic slows, say to 60 kmh. Some cars pull over at the shops. Some cars turn off to the suburbs. Others continue through the town, and rejoin the freeway, and continue at 100 kmh.

No analogy is exact, but the cars on the freeway are something like light in a vacuum. In a vacuum, light travels at a steady speed and in a straight line. Entering the town is something like light encountering matter; the speed limit decreases. When light encounters matter, it slows down. The change in speed is related to the refractive index and the phenomenon of *refraction*. Cars stopping are something like the *absorption* of light by matter; those vehicles are lost from the traffic flow. Some cars turn aside, just as the *scattering* of light diverts it from its original trajectory. Some cars may even make a U-turn and head back down the freeway along the direction they just came. This is something like the *reflection* of light. You can't make a U-turn on the freeway. The cars that continue on the freeway beyond the town are like the light that passes right through the matter, like the *transmission* of light. No analogy is exact, but the cars in some ways mimic the actions of photons.

So the interaction of light with matter includes reflection, refraction, absorption, scattering and transmission. Often, more than one interaction occurs at the same time. For example, when sunlight falls on a glass window, some is reflected back outside, some is scattered in sundry directions, some is absorbed in the glass, but most is transmitted.

The chapter is arranged like this. First we will look at a static electric field in the vacuum and in matter. This will introduce the key concept of the *dielectric function*. Light, as we saw in Chapter 4, is an interplay between ever changing electric and magnetic fields. So we move from a static electric field to a dynamic one. We extend Maxwell's equations, which we met in Chapter 4 for the case of a vacuum, to the case of matter. We look at the mathematical and physical significance of the new terms that appear in matter that were absent in the vacuum. We apply these equations to prototypical matter represented by a harmonic oscillator. We look then at the application to insulators and conductors. We conclude by considering what happens to light at an interface.

Learning goals

After studying this chapter you should be able to:

- appreciate the significance of the dielectric function,
- explain how Maxwell's equations change in going from a vacuum into matter and the mathematical and physical meaning of the new terms introduced,
- give an account of the interaction of light and matter using the model of a harmonic oscillator,
- discuss the propagation of light in insulators and conductors,
- give a qualitative and quantitative account of the reflection of light at the interface between two materials.

6.1 A static electric field in a region of space

In discussing the electric field in a region of space, terms such as 'dielectric', 'permittivity' and 'susceptibility' are used. This is so whether the region of space contains matter or is empty. I will now define the terms that I will use henceforth to talk about the electric field in a medium. I will start with the simplest medium – the vacuum – then go on to matter.

6.1.1 Vacuum

The electric constant, ϵ_0, is a fundamental physical constant. We have met the electric constant in Chapter 4 and it was defined in Equation (4.10). To recapitulate, it may be defined in terms of two other fundamental physical constants – the magnetic constant, μ_0, and the speed of light in the vacuum, c:

$$\epsilon_0 \equiv \frac{1}{\mu_0 c^2}. \tag{6.1}$$

This definition is a consequence of the profound relation between electricity and magnetism, birthing an electromagnetic wave, travelling always at lightspeed in a vacuum. Since the values for μ_0 and c are defined in the SI system, the value of ϵ_0 is also defined in the SI system.

$$\epsilon_0 = 8.854\,187\,817\cdots \times 10^{-12}\ \mathrm{F\,m^{-1}}. \tag{6.2}$$

The digits continuing does not indicate that the value is inexact but that it is irrational.

The units in Equation (6.2) are farads per metre, so ϵ_0 may be interpreted as connecting an electrical quantity with a mechanical quantity. The same conclusion may be drawn from inspecting Coulomb's law (arranged here to make ϵ_0 the subject),

$$\epsilon_0 = \frac{qQ}{4\pi r^2 F}. \tag{6.3}$$

The numerator involves electrical quantities, the two charges q and Q. The denominator involves mechanical quantities, the separation r and force F.

Permittivity The term 'permittivity' comes from the concept that a medium 'permits' an electric field to exist in it. So ϵ_0 is sometimes referred to as the 'vacuum permittivity', 'absolute permittivity', 'permittivity of free space', 'electric permittivity of free space'. I mention this in case you come across this in other contexts, but I will not use this terminology further.

Dielectric constant The term 'dielectric' derives from 'dia', meaning across, added to 'electric', and so from the concept that the electric field is found across a region of space. In the context of this terminology, ϵ_0 is sometimes referred to as the 'dielectric constant of free space' or 'dielectric constant of a vacuum'.

6.1.2 Matter

Let's rearrange Coulomb's law in a vacuum (Equation 6.3) to show explicitly the value of the electric field E. The electric field is defined to be the force per unit charge.

$$\epsilon_0 \frac{F}{q} = \epsilon_0 E = \frac{Q}{4\pi r^2}. \tag{6.4}$$

Assuming r, the position we are measuring the field, is not changing, the quantity ϵ_0 can then be identified as relating the charge Q to the electric field E.

We will now consider what happens when the region of space between the charges q and Q is not empty, but contains some matter. We will start with the least complicated matter: matter that fills the space uniformly, that is uniform throughout (homogeneous), whose properties are the same in all directions (isotropic), and that responds directly (linearly) to the electric field E and not to higher-order terms such as E^2 (which would be a nonlinear response). In general, when some matter is introduced into the space between the two charges, the electric field produced will be different from the electric field produced in the vacuum. We quantify this by modifying Equation (6.4). We may do this in two ways. We may look at the absolute change in the relation between Q and E and write

$$\epsilon E = \frac{Q}{4\pi r^2}. \tag{6.5}$$

Alternatively, we may express the change relative to the vacuum,

$$\epsilon_r \epsilon_0 E = \frac{Q}{4\pi r^2}. \tag{6.6}$$

The connection between the two newly introduced symbols is

$$\epsilon = \epsilon_r \epsilon_0. \tag{6.7}$$

What shall we call these new quantities, ϵ and ϵ_r? This requires some care. Often they are called the 'dielectric constant' and the 'relative dielectric constant'. This is because, under fixed conditions, such as constant temperature and pressure, the relation between E and Q in a given material is fixed. So the 'dielectric constant' can be considered to be a materials parameter, as density or heat capacity may be. The problem is, if the experimental conditions change, so does the value of the so-called constant. For example, the value may change with temperature and we may explicitly write ϵ as a function of temperature T as $\epsilon(T)$. It is reasonable to use the expression 'dielectric constant' provided it is understood we are referring to fixed conditions (usually room temperature and pressure).

I might clarify, the 'dielectric constant' of the vacuum, ϵ_0, is indeed a fixed value. It is defined by Equation (6.1) and does not vary with experimental conditions. For the vacuum $\epsilon \equiv \epsilon_0$ and $\epsilon_r \equiv 1$.

A couple more points of terminology. For many practical applications, materials with high values of ϵ_r are employed. It may be seen from Equation (6.6) that the effect of this, relative to the vacuum, is to increase the charge for a given electric field (so increasing the electrical capacitance), or, alternatively, to decrease the electric field in the material for a given charge. Materials with a high value of ϵ_r are often referred to as *dielectrics*, although of course many materials will change the relation of charge to field and so in that sense display dielectric behaviour. For a given material, at a high enough electric field, *dielectric breakdown* occurs, when the material can no longer sustain the charge separation across it, the insulating behaviour is lost, and a current flows. For the practical storage of electric charge in capacitors, the best materials to separate the charges have both high dielectric values and high dielectric breakdown fields.

To reiterate, to describe the dielectric nature of a material, the term 'dielectric constant' is often used, but the dielectric properties vary with such factors as temperature. When we move away from the static electric field that we have been considering in this section to the changing electric field that we will consider in the next section, the terminology 'dielectric constant' is rather dangerous. The dielectric properties of almost all materials change with frequency. So we best refer to the *dielectric function* and write it explicitly as a function of frequency or of angular frequency, $\epsilon(f)$ or $\epsilon(\omega)$. Likewise, the *relative dielectric function* is written as $\epsilon_r(f)$ or $\epsilon_r(\omega)$.

6.2 Electromagnetic waves in matter

Chapter 4 was all about electromagnetic waves in a vacuum. You may wish to review it before you read on. Key equations from that chapter gave the electric field in the vacuum and the magnetic field in a vacuum. Both equations had the same form. Here I will only repeat the equation for the electric field (Equation 4.12):

$$\nabla^2 \mathbf{E} = \mu_0 \epsilon_0 \frac{\partial^2 \mathbf{E}}{\partial t^2} . \qquad (6.8)$$

Now we will see how this equation is modified in matter. New terms are introduced. We will first look at the physical meaning, then, more abstractly, at the mathematical interpretation of these.

6.2.1 Additional terms in Maxwell's equations – physical meaning

In matter, Equation (6.8) is modified to

$$\nabla^2 \mathbf{E} = \mu_0 \sigma_{\text{MATTER}} \frac{\partial \mathbf{E}}{\partial t} + \mu_0 \epsilon_0 \chi \frac{\partial^2 \mathbf{E}}{\partial t^2} + \mu_0 \epsilon_0 \frac{\partial^2 \mathbf{E}}{\partial t^2} . \tag{6.9}$$

Let's look at the terms on the right-hand side of Equation (6.9). They contain the common factor μ_0. Leaving that aside, the physical significance of the three terms is

- $\sigma_{\text{MATTER}} \partial \mathbf{E}/\partial t$

  The quantity $\sigma_{\text{MATTER}} \mathbf{E}$ is the current density, usually denoted $\mathbf{J}$. Here we consider the current density to be directly proportional to the electric field $\mathbf{E}$. (In general things may be more complicated; the current density may not be in the same direction as the electric field, or the current density may not scale directly with the electric field.) The constant of proportionality is written here as σ_{MATTER} and is often called the *conductivity*; we will carefully define it in a moment. We assume that σ_{MATTER} does not vary in time, so that the rate of change of the current density, $\partial \mathbf{J}/\partial t = \sigma_{\text{MATTER}} \partial \mathbf{E}/\partial t$. So this term is simply the rate of change of current density.

- $\epsilon_0 \chi \partial^2 \mathbf{E}/\partial t^2$

  The quantity $\epsilon_0 \chi \mathbf{E}$ is analogous to the quantity $\epsilon_0 \mathbf{E}$ in the vacuum, the only difference being the factor χ to account for the presence of matter. (We assume here that the influence of the matter is directly proportional to the electric field, although in general it may be more complicated.) The quantity χ is known as the *electric susceptibility*. The combination $\epsilon_0 \chi$ is known as the *polarisability*. The combination $\epsilon_0 \chi \mathbf{E}$ physically corresponds to the electric dipole moment per unit volume in the matter. (This is also known as *electric polarisation*, and denoted $\mathbf{P}$, but I will try to limit my use of this term to minimise possible confusion with the polarisation of light.) Assuming that the electric dipole moment per unit volume changes directly in proportion to the electric field, the term $\epsilon_0 \chi$ can be brought out of the differentiation in $\partial^2 \epsilon_0 \chi \mathbf{E}/\partial t^2$. Hence the term $\epsilon_0 \chi \partial^2 \mathbf{E}/\partial t^2$ represents the second time derivative of the dipole moment per unit volume.

- $\epsilon_0 \partial^2 \mathbf{E}/\partial t^2$

  The first part of the term, ϵ_0, is simply the electric constant and the second part is simply the second time derivative of the electric field. This term is always present, even in the absence of matter. It may be interpreted as being related to an equivalent electric dipole moment per unit volume of the vacuum.

6.2.2 Additional terms in Maxwell's equations – mathematical relations

Let's now take a moment to consider the mathematical form of Equation (6.9), quite apart from its physical meaning. Let's consider the case of an electric field that varies sinusoidally with time. I will show we can write the equation either as

$$\nabla^2 \mathbf{E} = \mu_0 \tilde{\sigma} \frac{\partial \mathbf{E}}{\partial t} \tag{6.10}$$

6.2 Electromagnetic waves in matter

or as

$$\nabla^2 \mathbf{E} = \mu_0 \tilde{\epsilon} \frac{\partial^2 \mathbf{E}}{\partial t^2} . \quad (6.11)$$

Here the newly introduced term $\tilde{\sigma}$ is the *effective conductivity* and the newly introduced term $\tilde{\epsilon}$ is the *effective dielectric function*. They may be complex numbers, and I have introduced the tilde ($\tilde{\ }$) to admit this possibility.

If, as I am assuming, the electric field varies sinusoidally with time, the first and second derivatives are simply related. This allows us to eliminate one time derivative in favour of the other. If we write the time dependence of $\mathbf{E}$ as $\mathbf{E} = \mathbf{E}_0 \exp[-i\omega t]$, then $\partial \mathbf{E}/\partial t = -i\omega \mathbf{E}_0 \exp[-i\omega t] = -i\omega \mathbf{E}$. Differentiating again with respect to time, $\partial^2 \mathbf{E}/\partial t^2 = -i\omega \times (-i\omega \mathbf{E}_0 \exp[-i\omega t]) = -\omega^2 \mathbf{E}$. Every differentiation with respect to time is equivalent to multiplying by $-i\omega$. Putting this result in Equation (6.9) and Equation (6.10) yields

$$\mu_0 \tilde{\sigma}(-i\omega \mathbf{E}) = \mu_0 \sigma_{\text{MATTER}}(-i\omega \mathbf{E}) + \mu_0 \epsilon_0 \chi(-\omega^2 \mathbf{E}) + \mu_0 \epsilon_0 (-\omega^2 \mathbf{E}) , \quad (6.12)$$

which, on cancelling $-i\mu_0 \omega \mathbf{E}$ becomes

$$\tilde{\sigma} = \sigma_{\text{MATTER}} - \omega(\epsilon_0 \chi + \epsilon_0)i . \quad (6.13)$$

In the same way, putting the explicit time derivatives of the electric field in Equation (6.9) and Equation (6.11) gives

$$\mu_0 \tilde{\epsilon}(-\omega^2 \mathbf{E}) = \mu_0 \sigma_{\text{MATTER}}(-i\omega \mathbf{E}) + \mu_0 \epsilon_0 \chi(-\omega^2 \mathbf{E}) + \mu_0 \epsilon_0 (-\omega^2 \mathbf{E}) , \quad (6.14)$$

which, on cancelling $-\mu_0 \omega^2 \mathbf{E}$ gives

$$\tilde{\epsilon} = \epsilon_0 \chi + \epsilon_0 + \frac{\sigma_{\text{MATTER}}}{\omega} i . \quad (6.15)$$

A direct comparison of Equations (6.10) and (6.11) reveals

$$\mu_0 \tilde{\sigma}(-i\omega)\mathbf{E} = \mu_0 \tilde{\epsilon}(-\omega^2)\mathbf{E} , \quad (6.16)$$

and on cancelling $-\mu_0 \omega \mathbf{E}$

$$i\tilde{\sigma} = \omega \tilde{\epsilon} . \quad (6.17)$$

As we have seen in Section 2.2, we can separate the real and imaginary parts of the complex numbers. Here I will use the subscript 1 to denote the real part and the subscript 2 to denote the imaginary part.

$$\tilde{\sigma} = \sigma_1 + i\sigma_2. \quad (6.18)$$

Likewise,
$$\tilde{\epsilon} = \epsilon_1 + i\epsilon_2. \tag{6.19}$$

Using Equations (6.17), (6.18) and (6.19) allows us to write out explicit expressions for the real and imaginary parts of the effective conductivity and effective dielectric function in terms of the real and imaginary part of the other quantity.

$$\sigma_1 = \omega\epsilon_2. \tag{6.20}$$

$$\sigma_2 = -\omega\epsilon_1. \tag{6.21}$$

$$\epsilon_1 = -\frac{\sigma_2}{\omega}. \tag{6.22}$$

$$\epsilon_2 = \frac{\sigma_1}{\omega}. \tag{6.23}$$

6.3 Interaction of light and matter – harmonic oscillator model

We have seen in Section 6.2.1 that when electromagnetic waves pass through matter, two additional terms are introduced into Maxwell's equations, one related to the electric current density and one related to the electric dipole moment per unit volume.

We then saw in Section 6.2.2 that, for a sinusoidally varying electric field, the two terms are interchangeable; so the whole equation may be expressed in terms either of an effective conductivity, or an effective dielectric function, at our convenience.

Now let us apply these ideas to a particular physical system, the harmonic oscillator. We will see that we can analyse the harmonic oscillator either from the perspective of the dielectric function or from the perspective of the conductivity.

We are interested in the harmonic oscillator not only for its own sake but also as it serves as a model for many physical systems. We have previously seen that light can be considered as a harmonic oscillation; we will treat matter in a cognate way.

We will begin by writing down the equation of motion. We will consider motion in a single direction, which we shall call x. The acceleration will be denoted $\ddot{x}$. On one side of the equation will be the mass times the acceleration, $m\ddot{x}$. On the other side of the equation is the net force. There are three forces:

The driving force We assume a particle of charge q is subject to an electric field E. The electrical force on the particle is the product of the charge and the field, qE.

The damping force We assume the particle suffers a *damping force* proportional to its speed. We may write the damping force as $-\alpha\dot{x}$. The negative sign indicates that the force is in the opposite direction to the velocity, $\dot{x}$. It is convenient later to cancel out the mass, so rather than α we will write instead $m\gamma$ so the damping force is $-m\gamma\dot{x}$. Bearing in mind that force has dimensions of mass times acceleration, the quantity $\gamma\dot{x}$ must be an acceleration, so γ must be an

6.3 Interaction of light and matter – harmonic oscillator model

inverse time, or frequency. Hence another way to write the damping force is to use a characteristic damping time $\tau = 1/\gamma$.

The restoring force We assume the particle is pulled back, or restored, to its position by a force that is proportional to its displacement from equilibrium. Assuming we start measuring x from the equilibrium position, we can write the restoring force as $-\beta x$. The negative sign indicates that the force acts to return the particle to the equilibrium position. So, if the particle moves in the positive x direction, the force is in the negative x direction, and vice versa. Again, because it is convenient later to divide by m, we introduce m in the way we write the force constant β; again, the final expression has to correspond to a mass times an acceleration, so, given that x is involved, we have to divide twice by time, or, equivalently, multiply twice by frequency to obtain the correct dimensions. I will choose to multiply twice by (angular) frequency ω_0, and so write the restoring force as $-m\omega_0^2 x$.

In summary, we are considering a driven (E), damped (γ), harmonic (ω_0) oscillator. The equation of motion is

$$m\ddot{x} = qE - m\gamma\dot{x} - m\omega_0^2 x. \qquad (6.24)$$

We may express this more elegantly by leaving the driving force on the right-hand side, moving the restoring and damping forces to the left-hand side, and dividing by the mass:

$$\ddot{x} + \gamma\dot{x} + \omega_0^2 x = \frac{qE}{m}. \qquad (6.25)$$

The equation of motion is a rather general second-order differential equation. On the left-hand side, it contains terms in the second, first and zeroth time derivatives.

The solution of the equation of motion of the driven, damped harmonic oscillator depends in detail on how the right-hand side, the driving term, varies with time. We will make the assumption here that the motion has been established by an electric field that varies sinusoidally with time and that the position of the particle varies sinusoidally in the same way. Specifically, we will consider the position of the particle to vary as $x = x_0 \exp[-i\omega t]$. (Here x_0 represents the amplitude of motion, not the position at $t = 0$.) Then $\dot{x} = -i\omega x_0 \exp[-i\omega t] = -i\omega x$ and $\ddot{x} = -i\omega\dot{x} = -\omega^2 x$. Let's tabulate the relationships between the zeroth, first and second time derivatives of the position.

Table 6.1 Relationship of quantities in the harmonic oscillator model

Position	x	x	$x_0 \exp[-i\omega t]$	x	$+i\dfrac{v}{\omega}$	$-\dfrac{a}{\omega^2}$
Velocity	$\dot{x}$	v	$-i\omega x_0 \exp[-i\omega t]$	$-i\omega x$	v	$+i\dfrac{a}{\omega}$
Acceleration	$\ddot{x}$	a	$-\omega^2 x_0 \exp[-i\omega t]$	$-\omega^2 x$	$-i\omega v$	a

We can thus eliminate the terms in velocity and acceleration in the equation of motion in favour of position to obtain

$$-\omega^2 x - i\omega\gamma x + \omega_0^2 x = \frac{qE}{m}, \qquad (6.26)$$

or, making x the subject of the equation,

$$x = \frac{\frac{qE}{m}}{\omega_0^2 - \omega^2 - i\omega\gamma}. \qquad (6.27)$$

Alternatively, we can eliminate the terms in position and acceleration in the equation of motion in favour of velocity to obtain

$$-i\omega\dot{x} + \gamma\dot{x} + i\frac{\omega_0^2}{\omega}\dot{x} = \frac{qE}{m}, \qquad (6.28)$$

or

$$v = \dot{x} = \frac{\frac{qE}{m}}{\gamma + i\left(\frac{\omega_0^2}{\omega} - \omega\right)}. \qquad (6.29)$$

We can now apply this model to the electric field equation, Equation (6.9), either from the point of view of the conductivity, Equation (6.10), or from the point of view of the dielectric function, Equation (6.11).

Harmonic oscillator: dielectric function perspective

The dipole moment produced by a charge q moving a distance x, presuming an equal and opposite charge remains fixed at the original position, is given by qx. Assuming we have n mobile charges per unit volume, the dipole moment per unit volume is

$$nqx = nq\frac{\frac{qE}{m}}{\omega_0^2 - \omega^2 - i\omega\gamma} = \frac{\epsilon_0 \omega_p^2 E}{\omega_0^2 - \omega^2 - i\omega\gamma}, \qquad (6.30)$$

where I have introduced for convenience a term known as the *plasma frequency*,

$$\omega_p^2 \equiv \frac{nq^2}{\epsilon_0 m}. \qquad (6.31)$$

I will say more about the plasma frequency shortly. Recalling that the quantity $\epsilon_0 \chi E$ is the dipole moment per unit volume, we have

$$\epsilon_0 \chi = \frac{\epsilon_0 \omega_p^2}{\omega_0^2 - \omega^2 - i\omega\gamma}. \qquad (6.32)$$

6.3 Interaction of light and matter – harmonic oscillator model

So then, from Equation (6.15),

$$\tilde{\epsilon} = \epsilon_0 \chi + \epsilon_0 + \frac{\sigma_{\text{MATTER}}}{\omega} i = \frac{\epsilon_0 \omega_p^2}{\omega_0^2 - \omega^2 - i\omega\gamma} + \epsilon_0 + 0, \quad (6.33)$$

or

$$\tilde{\epsilon} = \epsilon_0 \left[\frac{\omega_p^2}{\omega_0^2 - \omega^2 - i\omega\gamma} + 1 \right]. \quad (6.34)$$

By first multiplying top and bottom by the complex conjugate of the denominator, $\omega_0^2 - \omega^2 + i\omega\gamma$, we may separate the real and imaginary parts of the effective dielectric function:

$$\epsilon_1 = \epsilon_0 \left[\frac{\omega_p^2 (\omega_0^2 - \omega^2)}{(\omega_0^2 - \omega^2)^2 + \omega^2\gamma^2} + 1 \right], \quad (6.35)$$

$$\epsilon_2 = \frac{\epsilon_0 \omega_p^2 \omega \gamma}{(\omega_0^2 - \omega^2)^2 + \omega^2\gamma^2}. \quad (6.36)$$

The real part of the dielectric function is shown in Figure 6.1.

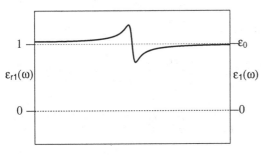

Figure 6.1 Real part of the dielectric function for the harmonic oscillator. The graph is scaled so that an angular frequency range ω: 0–10 corresponds to $\omega_0 = 5$, $\omega_p = 1$, $\gamma = 2$.

The imaginary part of the dielectric function is shown in Figure 6.2.

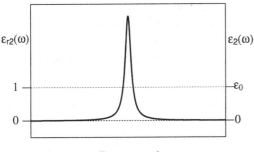

Figure 6.2 Imaginary part of the dielectric function for the harmonic oscillator. The graph is scaled so that an angular frequency range ω: 0–10 corresponds to $\omega_0 = 5$, $\omega_p = 1$, $\gamma = 2$.

Harmonic oscillator: conductivity perspective

The current density produced by a charge density nq moving at speed v is

$$nqv = nq\frac{\frac{qE}{m}}{\gamma + i\left(\frac{\omega_0^2}{\omega} - \omega\right)} = \frac{\epsilon_0\omega_p^2\omega E}{\omega\gamma + i\left(\omega_0^2 - \omega^2\right)}. \tag{6.37}$$

Recalling that the current density is $\sigma_{\text{MATTER}}E$, we may write

$$\sigma_{\text{MATTER}} = \frac{\epsilon_0\omega_p^2\omega}{\omega\gamma + i\left(\omega_0^2 - \omega^2\right)}. \tag{6.38}$$

Hence, from Equation (6.13),

$$\tilde{\sigma} = \sigma_{\text{MATTER}} - \omega(\epsilon_0\chi + \epsilon_0)i = \frac{\epsilon_0\omega_p^2\omega}{\omega\gamma + i\left(\omega_0^2 - \omega^2\right)} - \omega(0 + \epsilon_0)i. \tag{6.39}$$

Thus

$$\tilde{\sigma} = \epsilon_0\omega\left[\frac{\omega_p^2}{\omega\gamma + i\left(\omega_0^2 - \omega^2\right)} - i\right]. \tag{6.40}$$

Realising the denominator,

$$\sigma_1 = \frac{\epsilon_0\omega_p^2\omega^2\gamma}{(\omega_0^2 - \omega^2)^2 + \omega^2\gamma^2}, \tag{6.41}$$

$$\sigma_2 = -\epsilon_0\omega\left[\frac{(\omega_0^2 - \omega^2)\omega_p^2}{(\omega_0^2 - \omega^2)^2 + \omega^2\gamma^2} + 1\right]. \tag{6.42}$$

6.3.1 Harmonic oscillator: without harmony – the conductor

We may model a *conductor* as a harmonic oscillator without harmony. By this I mean we ignore the spring constant. Instead of picturing an electron as bound to a nucleus by a little spring, we picture it as being unbound. It is not entirely free, though – it is still subject to the damping force. Mathematically, we use exactly the same analysis as for the harmonic oscillator, but simply set ω_0 to zero.

The effective dielectric function for the conductor follows directly from Equation (6.34) with $\omega_0 = 0$:

$$\tilde{\epsilon} = \epsilon_0\left[1 - \frac{\omega_p^2}{\omega^2 + i\omega\gamma}\right]. \tag{6.43}$$

6.3 Interaction of light and matter – harmonic oscillator model

The real and imaginary parts follow from Equations (6.35) and (6.36) with $\omega_0 = 0$:

$$\epsilon_1 = \epsilon_0 \left[1 - \frac{\omega_p^2}{\omega^2 + \gamma^2}\right], \tag{6.44}$$

$$\epsilon_2 = \frac{\epsilon_0 \omega_p^2 \gamma}{\omega^3 + \omega\gamma^2}. \tag{6.45}$$

The real part of the dielectric function is shown in Figure 6.3. The imaginary part of

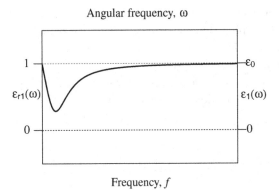

Figure 6.3 Real part of the dielectric function for a conductor. The graph is scaled so that an angular frequency range ω: 0–10 corresponds to $\omega_p = 1$, $\gamma = 2$. (For the conductor, $\omega_0 = 0$.)

the dielectric function is shown in Figure 6.4.

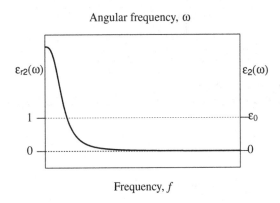

Figure 6.4 Imaginary part of the dielectric function for a conductor. The graph is scaled so that an angular frequency range ω: 0–10 corresponds to $\omega_p = 1$, $\gamma = 2$. (For the conductor, $\omega_0 = 0$.)

The effective conductivity for the conductor follows directly from Equation (6.40) with $\omega_0 = 0$:

$$\tilde{\sigma} = \epsilon_0 \omega \left[\frac{\omega_p^2}{\omega\gamma - i\omega^2} - i\right]. \tag{6.46}$$

The real and imaginary parts follow from Equations (6.41) and (6.42) with $\omega_0 = 0$:

$$\sigma_1 = \frac{\epsilon_0 \omega_p^2 \gamma}{\omega^2 + \gamma^2}, \qquad (6.47)$$

$$\sigma_2 = \epsilon_0 \omega \left[\frac{\omega_p^2}{\omega^2 + \gamma^2} - 1 \right]. \qquad (6.48)$$

6.3.2 Harmonic oscillator: without harmony, or damping – the plasma

We may model a *plasma* as a harmonic oscillator without the spring force and without the damping force. In other words, the only force acting on the particle is the electric force provided by the driving field. Mathematically, we use exactly the same analysis as for the harmonic oscillator, but with both ω_0 and γ set to zero.

The effective dielectric function for the plasma follows directly from Equation (6.34) with $\omega_0 = 0$ and $\gamma = 0$ (or from Equation (6.43) with $\gamma = 0$):

$$\tilde{\epsilon} = \epsilon_0 \left[1 - \frac{\omega_p^2}{\omega^2} \right]. \qquad (6.49)$$

The real and imaginary parts follow from Equations (6.35) and (6.36) with $\omega_0 = 0$ and $\gamma = 0$ (or from Equations (6.44) and (6.45) with $\gamma = 0$):

$$\epsilon_1 = \epsilon_0 \left[1 - \frac{\omega_p^2}{\omega^2} \right], \qquad (6.50)$$

$$\epsilon_2 = 0. \qquad (6.51)$$

So the effective dielectric function is purely real.

The real part of the dielectric function is shown in Figure 6.5. The imaginary part of

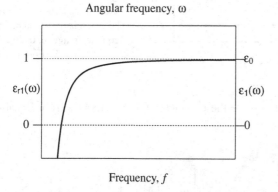

Figure 6.5 Real part of the dielectric function for a plasma. The graph is scaled so that an angular frequency range ω: 0–10 corresponds to $\omega_p = 1$. (For the plasma, $\omega_0 = 0$ and $\gamma = 0$.)

the dielectric function is shown in Figure 6.6.

6.3 Interaction of light and matter – harmonic oscillator model

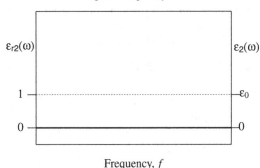

Figure 6.6 Imaginary part of the dielectric function for a plasma. It is identically zero. (For the plasma, $\omega_0 = 0$ and $\gamma = 0$.)

The effective conductivity for the plasma follows directly from Equation (6.40) with $\omega_0 = 0$ and $\gamma = 0$ (or from Equation (6.46) with $\gamma = 0$):

$$\tilde{\sigma} = \epsilon_0 \omega \left[\frac{\omega_p^2}{\omega^2} - 1 \right] i . \tag{6.52}$$

The real and imaginary parts follow from Equations (6.41) and (6.42) with $\omega_0 = 0$ and $\gamma = 0$ (or from Equations (6.47) and (6.48) with $\gamma = 0$):

$$\sigma_1 = 0 , \tag{6.53}$$

$$\sigma_2 = \epsilon_0 \omega \left[\frac{\omega_p^2}{\omega^2} - 1 \right] . \tag{6.54}$$

So the effective conductivity is purely imaginary.

6.3.3 Multiple harmonic oscillators

We can extend the model of the harmonic oscillator to multiple oscillators. Assuming the oscillators, labelled by the index i, all act independently, the total effect is

$$\tilde{\epsilon} = \epsilon_0 + \epsilon_0 \sum_i \frac{\frac{n_i q_i^2}{\epsilon_0 m_i}}{\omega_i^2 - \omega^2 - i\omega\gamma_i} . \tag{6.55}$$

6.3.4 Lattice harmonic oscillation

We can extend the model of the harmonic oscillator to a lattice, assuming the lattice as a whole responds as a single particle. Then

$$\tilde{\epsilon} = \epsilon_L(0) + \frac{s_L}{\omega_L^2 - \omega^2 - i\omega\gamma_L} , \tag{6.56}$$

where $\epsilon_L(0)$ is static dielectric function of the lattice vibrations, ω_L is the resonance frequency, s_L is the oscillator strength and γ_L is the damping constant.

6.4 Electromagnetic waves in an insulator

Let's now look in a little more detail at the insulator. We take the conductivity to be zero and the dielectric function to be real and so write

$$\nabla^2 \mathbf{E} = \mu_0 \epsilon(\omega) \frac{\partial^2 \mathbf{E}}{\partial t^2} \tag{6.57}$$

and

$$\nabla^2 \mathbf{B} = \mu_0 \epsilon(\omega) \frac{\partial^2 \mathbf{B}}{\partial t^2}. \tag{6.58}$$

Each of these equations is of the form of a wave equation, and the wavespeed is given in each case as

$$v = \frac{1}{\sqrt{\mu_0 \epsilon(\omega)}}. \tag{6.59}$$

If the wavespeed depends on the frequency, as is explicitly possible here, we say the material is *dispersive*. Likewise, if the wavespeed is the same for all frequencies, the material is called *non-dispersive*. This equation for wavespeed in matter differs only by a missing subscript and an added dependence on ω from the expression for the vacuum lightspeed

$$c = \frac{1}{\sqrt{\mu_0 \epsilon_0}}. \tag{6.60}$$

The ratio of these two speeds is defined to be the *refractive index*,

$$n(\omega) \equiv \frac{c}{v}. \tag{6.61}$$

Combining the last three equations, we may write

$$n(\omega) = \frac{c}{v} = \frac{\sqrt{\mu_0 \epsilon(\omega)}}{\sqrt{\mu_0 \epsilon_0}} = \sqrt{\frac{\epsilon(\omega)}{\epsilon_0}}. \tag{6.62}$$

This last ratio is directly related to the expression for the relative dielectric function, introduced in Equation (6.7), and now written explicitly as a function of ω:

$$n(\omega) = \sqrt{\epsilon_r(\omega)}. \tag{6.63}$$

We may rearrange the equation to make $\epsilon_r(\omega)$ the subject:

$$\epsilon_r(\omega) = n^2(\omega). \tag{6.64}$$

We may relate the refractive index to the angular wavenumber, k, by recalling that the wavespeed $v = \omega/k$. Substituting this into the definition of refractive index, Equation (6.61),

$$n(\omega) = k\frac{c}{\omega}. \tag{6.65}$$

Rearranging to make k the subject,

$$k = n(\omega)\frac{\omega}{c}. \tag{6.66}$$

The special circumstances in which this analysis and these results hold: for an uncharged, non-magnetic, insulating material for which ϵ is a real, scalar quantity varying only with ω. In this case, the solution to Maxwell's equations is a propagating electromagnetic wave. The amplitude, E, is the same as in a vacuum. The angular frequency, ω, is the same as in a vacuum. The angular wavenumber, k, differs from the vacuum value of ω/c by the factor of the refractive index, $n(\omega)$; consequently, the speed of the wave is reduced by a factor of $n(\omega)$. The refractive index is equal to the square root of the relative dielectric function, $\epsilon_r(\omega)$.

6.5 Electromagnetic waves in a conductor

Let's now look in a little more detail at the conductor. We will cast the initial equations in a form where σ_c and ϵ_c are real quantities (that may depend on ω) and are identified with the conductor through the subscript 'c':

$$\nabla^2 \mathbf{E} = \mu_0 \sigma_c(\omega)\frac{\partial \mathbf{E}}{\partial t} + \mu_0 \epsilon_c(\omega)\frac{\partial^2 \mathbf{E}}{\partial t^2} \tag{6.67}$$

and

$$\nabla^2 \mathbf{B} = \mu_0 \sigma_c(\omega)\frac{\partial \mathbf{B}}{\partial t} + \mu_0 \epsilon_c(\omega)\frac{\partial^2 \mathbf{B}}{\partial t^2}. \tag{6.68}$$

It is perhaps not so easy to see the solution to these equations, involving as they do the additional conductivity term, but let's assume a solution in the form of a wave equation. For simplicity, we will assume the wave is propagating in the z direction and write

$$E = E_0 \exp[i(\tilde{k}z - \omega t)]. \tag{6.69}$$

We will see this fits the bill, with the proviso that the angular wavenumber, $\tilde{k}$, is now permitted to be a complex number, as I have explicitly written it. Taking the first two partial derivatives with respect to space:

$$\frac{dE}{dz} = i\tilde{k}E_0 \exp[i(\tilde{k}z - \omega t)] = i\tilde{k}E, \tag{6.70}$$

$$\frac{d^2E}{dz^2} = (i\tilde{k})(i\tilde{k})E_0\exp[i(\tilde{k}z - \omega t)] = -\tilde{k}^2 E. \tag{6.71}$$

Take the first two partial derivatives with respect to time:

$$\frac{\partial E}{\partial t} = -i\omega E_0\exp[i(\tilde{k}z - \omega t)] = -i\omega E, \tag{6.72}$$

$$\frac{\partial^2 E}{\partial t^2} = (-i\omega)(-i\omega)E_0\exp[i(\tilde{k}z - \omega t)] = -\omega^2 E. \tag{6.73}$$

Feeding these back into Equation (6.67) yields

$$-\tilde{k}^2 E = \mu_0 \sigma_c(\omega)[-i\omega E] + \mu_0 \epsilon_c(\omega)[-\omega^2 E)], \tag{6.74}$$

which is simplified, on cancelling $-E$, to

$$\tilde{k}^2 = \mu_0 \epsilon_c(\omega)\omega^2 + \mu_0 \sigma_c(\omega)\omega i. \tag{6.75}$$

(The same equation can be obtained for the magnetic field equations.) The key point is that the right-hand side of this equation is explicitly a complex quantity, consisting of a real and an imaginary part, and so $\tilde{k}^2$ and in turn $\tilde{k}$ are also, in general, complex. The right-hand side of the equation also has an explicit dependence on ω. Hence $\tilde{k}^2$ and $\tilde{k}$ are also, in general, functions of ω.

Now let's reformulate the right-hand side by taking out the common factor of ω^2 and replacing μ_0 with $1/c^2\epsilon_0$:

$$\tilde{k}^2 = \omega^2 \left(\mu_0 \epsilon_c(\omega) + \frac{\mu_0 \sigma_c(\omega)}{\omega}i\right) = \frac{\omega^2}{c^2}\left(\frac{\epsilon_c(\omega)}{\epsilon_0} + \frac{\sigma_c(\omega)}{\epsilon_0 \omega}i\right). \tag{6.76}$$

Taking the square root of each side of this equation gives an equation of the form of Equation (6.66),

$$\tilde{k} = \frac{\omega}{c}\tilde{n}(\omega), \tag{6.77}$$

where

$$\tilde{n}(\omega) = \sqrt{\frac{\epsilon_c(\omega)}{\epsilon_0} + \frac{\sigma_c(\omega)}{\epsilon_0 \omega}i}. \tag{6.78}$$

In analogy with Equation (6.61), where the ratio c/v appears, but using $\omega/\tilde{k}$ for v, we can identify the *complex refractive index* as

$$\tilde{n}(\omega) = \frac{c}{\omega}\tilde{k}. \tag{6.79}$$

This has the same form as the real Equation (6.66).

In analogy with Equation (6.64), we may now define a complex relative dielectric function:

$$\tilde{\epsilon}_r(\omega) \equiv \tilde{n}^2(\omega). \tag{6.80}$$

We make $\tilde{n}(\omega)$ the subject instead, analogous to Equation (6.63).

$$\tilde{n}(\omega) = \sqrt{\tilde{\epsilon}_r(\omega)}. \tag{6.81}$$

Let us spend a moment looking at the relationships between the real and imaginary components of $\tilde{n}(\omega)$ and $\tilde{\epsilon}_r(\omega)$ purely from a mathematical point of view, on the basis that we have two complex quantities, one the square of the other. After that, we will look further at the physical meaning of the terms. Let us write out explicitly the real and imaginary components of the refractive index and the dielectric function. For convenience, I will drop the explicit dependence on ω in the notation here.

$$\tilde{n} = n_1 + n_2 i. \tag{6.82}$$

$$\tilde{\epsilon}_r = \epsilon_{r1} + \epsilon_{r2} i. \tag{6.83}$$

The quantities n_1, n_2, ϵ_{r1} and ϵ_{r2} are all real. To go from $\tilde{n}$ to $\tilde{\epsilon}_r$ is relatively straightforward. Squaring the complex quantity $\tilde{n}$ gives

$$\tilde{n}^2 = (n_1 + n_2 i) \times (n_1 + n_2 i) = n_1^2 - n_2^2 + 2n_1 n_2 i. \tag{6.84}$$

Since $\tilde{\epsilon}_r = \tilde{n}^2$, we can directly write down the real and imaginary parts of $\tilde{\epsilon}_r$ as

$$\epsilon_{r1} = n_1^2 - n_2^2, \tag{6.85}$$

$$\epsilon_{r2} = 2n_1 n_2. \tag{6.86}$$

Example 6.1 Dependence of $\tilde{\epsilon}_r$ on $\tilde{n}$

What are the real and imaginary components of $\tilde{\epsilon}_r$ if (i) $\tilde{n}$ is purely real, (ii) $\tilde{n}$ is purely imaginary? (iii) What constraint on the real and imaginary components of $\tilde{n}$ is needed to ensure $\tilde{\epsilon}_r$ purely imaginary?

Solution

(i) If $\tilde{n}$ is purely real, the imaginary component $n_2 = 0$. Then $\epsilon_{r2} = 0$, and so $\tilde{\epsilon}_r$ is also purely real. We may write $\epsilon_{r1} = n_1^2$, echoing Equation (6.64).

(ii) On the other hand, if $\tilde{n}$ is purely imaginary, the real component $n_1 = 0$. In this case $\tilde{\epsilon}_r$ is still purely real, and has the value $\epsilon_{r1} = -n_2^2$. (This result may seem surprising, but squaring a purely imaginary number always yields a negative real number; the imaginary numbers were developed for this very property.)

(iii) To make $\tilde{\epsilon}_r$ purely imaginary requires $\epsilon_{r1} = n_1^2 - n_2^2$ to be zero. This requires $n_1 = \pm n_2$. (It may help you to picture lines at $\pm \pi/4$ to the real axis on the complex plane.)

We have the components of $\tilde{\epsilon}_r$ in terms of the components of $\tilde{n}$. How about the reverse, the components of $\tilde{n}$ in terms of the components of $\tilde{\epsilon}_r$? This is more difficult, as we are now taking the square root of a complex number. I won't give the working, but only the result here:

$$n_1 = \sqrt{\frac{1}{2}\sqrt{\sqrt{\epsilon_{r1}^2 + \epsilon_{r2}^2} + \epsilon_{r1}}}, \qquad (6.87)$$

$$n_2 = \sqrt{\frac{1}{2}\sqrt{\sqrt{\epsilon_{r1}^2 + \epsilon_{r2}^2} - \epsilon_{r1}}}. \qquad (6.88)$$

You may notice in the last two equations the expression $\sqrt{\epsilon_{r1}^2 + \epsilon_{r2}^2}$. This has a simple mathematical meaning. It is the square root of the sum of the squares of the real and imaginary components of a complex quantity. In general, this is the magnitude of the complex quantity, in this case $|\tilde{\epsilon}_r|$. So we may recast the previous two equations as

$$n_1 = \sqrt{\frac{|\tilde{\epsilon}_r| + \epsilon_{r1}}{2}}, \qquad (6.89)$$

$$n_2 = \sqrt{\frac{|\tilde{\epsilon}_r| - \epsilon_{r1}}{2}}. \qquad (6.90)$$

Exercise 6.1 Refractive index
If $\tilde{\epsilon}_r$ is purely real ($\epsilon_{r2} = 0$), what are the values of n_1 and n_2? What are the values of n_1 and n_2 if $\tilde{\epsilon}_r$ is purely imaginary? ∎

The relations just given between n_1, n_2, ϵ_{r1} and ϵ_{r2} are solely mathematical results based on the fact that one quantity, the dielectric function, may be written in terms of another entity, the refractive index, squared. Let us now return to the physics and look into the physical meaning of n_1, n_2, ϵ_{r1} and ϵ_{r2}. It is simplest to start with the dielectric function, then return to the refractive index. From Equations (6.78) and (6.80),

$$\tilde{\epsilon}_r(\omega) = \frac{\epsilon_c(\omega)}{\epsilon_0} + \frac{\sigma_c(\omega)}{\epsilon_0 \omega} i, \qquad (6.91)$$

or

$$\epsilon_{r1}(\omega) = \frac{\epsilon_c(\omega)}{\epsilon_0}, \qquad (6.92)$$

$$\epsilon_{r2}(\omega) = \frac{\sigma_c(\omega)}{\epsilon_0 \omega}. \qquad (6.93)$$

6.5 Electromagnetic waves in a conductor

So the real part of the complex dielectric function is related to the dielectric function as originally defined, whereas the complex part is related to the conductivity.

To obtain expressions for the real and imaginary parts of the refractive index, we can insert these results into Equations (6.87) and (6.88). The result is rather messy and may not give direct insight into the physical significance of $n_1(\omega)$ and $n_2(\omega)$. Let's try to get a grasp of their physical significance by first rewriting $\tilde{k}$ in terms of them, then substituting this expression in the equation of the electric wave. From Equation (6.77),

$$\tilde{k} = \tilde{n}(\omega)\frac{\omega}{c} = n_1(\omega)\frac{\omega}{c} + n_2(\omega)\frac{\omega}{c}i. \tag{6.94}$$

Now substituting this expression into Equation (6.69),

$$\begin{aligned}E &= E_0 \exp[i(\tilde{k}z - \omega t)] \\ &= E_0 \exp[i([n_1(\omega)\frac{\omega}{c} + n_2(\omega)\frac{\omega}{c}i]z - \omega t)] \\ &= E_0 \exp[i(n_1(\omega)\frac{\omega}{c}z - \omega t)] \exp[-n_2(\omega)\frac{\omega}{c}z].\end{aligned} \tag{6.95}$$

In the last step, the expression containing two imaginary terms, $i \times in_2(\omega)\omega/c$, becomes real (and negative, as it happens), and is separated out from the remaining terms, which remain imaginary.

Let's take a look at this last equation. The first part, $E_0 \exp[i(n_1(\omega)\omega z/z - \omega t)]$, is exactly the equation for a wave propagating in an insulator, or in a lossless medium. It is exactly the equation for a wave propagating in a vacuum with the wavespeed changed from c to $c/n_1(\omega)$. So the physical meaning of $n_1(\omega)$, the real part of the complex refractive index in a conductor, is the same as the (real) refractive index in an insulator. The second part, $\exp[-n_2(\omega)\omega z/c]$, represents an exponential decay in the conductor with characteristic length that depends on $n_2(\omega)$. For this reason, $n_2(\omega)$ is given the name *extinction coefficient*.

Often, we are concerned with the amplitude of the electric field, rather than the value of the electric field itself. The amplitude is proportional to the electric field squared. In this case, the exponential decay term is also squared, and becomes $\exp[-2n_2(\omega)\omega z/c]$. The amplitude of this exponential decay term may be written

$$A(z, \omega) = A_0 \exp[-2n_2(\omega)\frac{\omega}{c}z] = A_0 \exp[-\alpha(\omega)z], \tag{6.96}$$

where

$$\alpha(\omega) = 2n_2(\omega)\frac{\omega}{c} \tag{6.97}$$

is the *absorption coefficient*. The absorption coefficient has dimension of inverse length. The SI units for α are therefore m^{-1}. In practice, cm^{-1} are often used. A larger absorption coefficient corresponds to stronger absorption, to a more rapid decay in amplitude. We may express the absorption strength in the inverse way. We may ask, over what length does the amplitude decay by a factor of $1/e$? The answer is the value of z for which $\alpha(\omega)z = 1$. This is

$$\delta(\omega) = \frac{c}{2n_2(\omega)\omega}. \qquad (6.98)$$

The extinction coefficient, $n_2(\omega)$, the absorption coefficient, $\alpha(\omega)$, and the decay length, $\delta(\omega)$, are distinct but related ways of describing the attenuation of an electromagnetic wave as it travels through matter. In general, each of the three terms depends on the frequency of the electromagnetic wave.

Let me summarise our analysis for an uncharged, non-magnetic, conducting material in which the dielectric function is a scalar quantity varying only with ω. We have found that the angular wavenumber, $\tilde{k}$, the refractive index, $\tilde{n}$, and the dielectric function, $\tilde{\epsilon}$, are all complex quantities. The solution to Maxwell's equations is a propagating electromagnetic wave that decays with distance. The angular frequency, ω, is the same as in a vacuum. The speed of the wave is reduced by a factor of the real part of the refractive index. The magnitude of the electric field falls away exponentially with distance. The characteristic decay length is fixed by the imaginary part of the refractive index. The relative dielectric function is equal to the square of the refractive index. The real part of the dielectric function is related to the relative dielectric function, as in an insulator. The imaginary part of the dielectric function is related to the conductivity.

6.6 Electromagnetic waves at an interface

In our analysis so far, we have considered electromagnetic waves travelling in a uniform medium. In the vacuum, all electromagnetic waves travel at lightspeed. In an insulator, not much is different, except the wavespeed is now reduced, and possibly by different amounts at different frequencies. In a conductor, the situation is slightly different again. As well as being slowed, the waves are now attenuated exponentially with distance. The absorption coefficient, which describes the attenuation, in general depends on the frequency. Of the phenomena mentioned at the beginning of the chapter, so far we have only met 'absorption' and 'transmission'. What about 'reflection' and 'refraction'?

Reflection and refraction only occur at an interface; that is, when an electromagnetic wave leaves one medium and enters another. In contrast to uniform media, where light travels in a straight line, light now may be bent (refracted) or sent back the way it came (reflected).

We will look at the general case of light at an interface. We will start with the conceptually simple case of one medium, which we will call medium 1, that fills all of space for negative values of z and a second medium, which we will name medium 2, that fills all of space for positive values of z. The boundary or interface between these two media is at $z = 0$; in other words, the interface is the x-y plane.

For light in a vacuum striking the interface with a medium of refractive index $\tilde{n}(\omega)$, the reflectivity (also known as the power reflectance coefficient) is

$$R(\omega) = \left|\frac{1-\tilde{n}(\omega)}{1+\tilde{n}(\omega)}\right|^2 = \left|\frac{(n_1(\omega)-1)^2 + n_2(\omega)^2}{(n_1(\omega)+1)^2 + n_2(\omega)^2}\right|. \tag{6.99}$$

We can look at two extreme cases: a very non-conductive material, and a very conductive material.

6.6.1 Reflection from an insulator

In the case of the non-conductor (insulator) the conductivity can be taken to be zero. Therefore (from Equation (6.78), for example), $n_2(\omega) = 0$. The reflectivity equation is then

$$R(\omega) = \left|\frac{(n_1(\omega)-1)^2}{(n_1(\omega)+1)^2}\right| = \left(\frac{n_1(\omega)-1}{n_1(\omega)+1}\right)^2. \tag{6.100}$$

In the last step, we could drop the modulus sign that applies to complex numbers, since $n_1(\omega)$ is (by definition) real.

Example 6.2 Light reflected from silicon

Consider Si to have a real refractive index at 1 THz of 3.5. What amount of radiant power in a vacuum striking a surface of silicon will be reflected? How much will be transmitted?

Solution
Here the refractive index is purely real and specified at the single frequency of interest, so we may write $\tilde{n}(1\text{ THz}) = n_1 = 3.5$. The imaginary part is zero, $n_2 = 0$. Putting these values into Equation (6.99),

$$R(1\text{ THz}) = \left|\frac{(3.5-1)^2 + 0^2}{(3.5+1)^2 + 0^2}\right| = \left|\frac{2.5^2}{4.5^2}\right| = 0.31. \tag{6.101}$$

So 31% of the incident power is reflected. The rest of the power is transmitted:

$$T = 1 - R = 1 - 0.31 = 0.69. \tag{6.102}$$

So 69% of the incident power is transmitted.

6.6.2 Reflection from a conductor at low frequency

For a conductor, at low enough freqency, the conductivity terms will dominate the dielectric terms. In this regime, the real and the imaginary components of the refractive index approach each other and have the very large value

$$n_1(\omega \to 0) \sim n_2(\omega \to 0) \sim \sqrt{\frac{\sigma_0}{2\epsilon_0\omega}} = \sqrt{\frac{\sigma_0}{4\pi\epsilon_0 f}}. \tag{6.103}$$

Example 6.3 Refractive index of copper

Copper is a very good conductor. Its conductivity is about 5.9×10^7 Ω^{-1}-m^{-1}. Approximately what are the real and imaginary parts of the refractive index of copper at (i) 0.1 THz, (ii) 1 THz and (iii) 10 THz?

Solution

We use Equation (6.103) with the values $\sigma_0 = 5.9 \times 10^7$ Ω^{-1}-m^{-1} and f successively 10^{11}, 10^{12} and 10^{13} Hz to obtain n_1(Copper, 0.1 THz) $\sim$ 2300, n_1(Copper, 1 THz) $\sim$ 730, n_1(Copper, 10 THz) $\sim$ 230. Notice that all these values are much greater than 1. The imaginary parts are the same as the real parts.

Putting $n_1(\omega \to 0) = n_2(\omega \to 0) \gg 1$ into Equation (6.99) gives

$$R(\omega \to 0) = \frac{(n_1(\omega) - 1)^2 + n_1(\omega)^2}{(n_1(\omega) + 1)^2 + n_1(\omega)^2}. \tag{6.104}$$

To a first approximation, if $n_1(\omega) \gg 1$, both numerator and denominator are about the same, about $2n_1(\omega)^2$; so the ratio is 1; in other words, all light is reflected. This means in essence all conductors are perfect reflectors for light of low enough frequency. We obtain a better approximation by expanding out the squared terms

$$\begin{aligned}R(\omega \to 0) &= \frac{n_1(\omega)^2 - 2n_1(\omega) + 1 + n_1(\omega)^2}{n_1(\omega)^2 + 2n_1(\omega) + 1 + n_1(\omega)^2} \\ &\sim \frac{2n_1(\omega)^2 - 2n_1(\omega)}{2n_1(\omega)^2 + 2n_1(\omega)} \\ &= \frac{n_1(\omega) - 1}{n_1(\omega) + 1} \\ &\sim 1 - \frac{2}{n_1(\omega)}. \end{aligned} \tag{6.105}$$

We may now insert in this result the expression for $n_1(\omega)$ we obtained previously (Equation 6.103):

$$R(\omega \to 0) \sim 1 - 2\sqrt{\frac{2\epsilon_0 \omega}{\sigma_0}} = 1 - \sqrt{\frac{16\pi\epsilon_0 f}{\sigma_0}}. \tag{6.106}$$

This result is known as the Hagen-Rubens relation. It says that the difference from perfect reflectivity increases with the square root of the frequency.

We may give a numerical expression for R which is useful in calculations. If the frequency is expressed in units of terahertz and the conductivity is expressed in the SI units of inverse ohm – inverse metre, we have

$$R(\omega \to 0) \sim 1 - 21.096 \sqrt{\frac{f \text{ [in terahertz]}}{\sigma_0 \text{ [in } \Omega^{-1}\text{-m}^{-1}\text{]}}}. \tag{6.107}$$

6.6 Electromagnetic waves at an interface

Example 6.4 Reflectivity of copper

For light in a vacuum of frequencies (i) 0.1 THz, (ii) 1 THz and (iii) 10 THz, approximately what fraction is reflected at normal incidence from copper? Take the conductivity of copper to be 5.9×10^7 Ω^{-1}-m^{-1}.

Solution

We use Equation (6.107) with the numerical values $\sigma_0 = 5.9 \times 10^7$ and f successively 0.1, 1 and 10 to obtain R(Copper, 0.1 THz) $\sim$ 99.9%, R(Copper, 1 THz) $\sim$ 99.7%, R(Copper, 10 THz) $\sim$ 99.1%. Notice that these values are all within 1% of 1.

We may ask the question, does all the light reflect from the conductor, does none get through? As the numerical examples have shown, almost all the light is reflected, but not quite all. Some penetrates a short distance into the conductor.

To calculate the penetration depth, we refer back to Equation (6.95). There appeared a term $\exp[-n_2(\omega)\frac{\omega}{c}z]$, which relates to the decay of the wave within the conductor. We may establish a characteristic decay length δ and write more compactly $\exp[-z/\delta]$, where

$$\delta \equiv \frac{c}{n_2(\omega)\omega}. \tag{6.108}$$

(Don't confuse this length with the decay length defined in Equation (6.98). That refers to the decay in intensity. This refers to the decay in field.) In conductors, as we shall see, this distance is rather small, and is referred to as the *skin depth*. Now, inserting the refractive index expression for a conductor, Equation (6.103), we may write

$$\delta = \frac{c}{\omega}\sqrt{\frac{2\epsilon_0\omega}{\sigma_0}} = \sqrt{\frac{2\epsilon_0 c^2}{\sigma_0 \omega}} = \sqrt{\frac{2}{\mu_0 \sigma_0 \omega}} = \sqrt{\frac{\epsilon_0 c^2}{\pi \sigma_0 f}} = \sqrt{\frac{1}{\pi \mu_0 \sigma_0 f}}. \tag{6.109}$$

(In obtaining the various expressions, I have used $c^2 = 1/\mu_0\epsilon_0$ and $\omega = 2\pi f$.) Here is a numerical expression for δ that may be useful in calculations. If the frequency is expressed in units of terahertz and the conductivity is expressed in the SI units of inverse ohm-inverse metre, we have

$$\delta \text{ [in nanometres]} = \frac{503292}{\sqrt{\sigma_0 \text{ [in } \Omega^{-1}\text{-m}^{-1}] f \text{ [in terahertz]}}}. \tag{6.110}$$

Example 6.5 Skin depth of copper

How far does terahertz radiation penetrate into copper ($\sigma_0 = 5.9 \times 10^7$ Ω^{-1}-m^{-1})? Give an answer for frequencies of 0.1, 1 and 10 THz, and compare with the vacuum wavelengths in each case.

Solution

We use Equation (6.110) with the numerical values $\sigma_0 = 5.9 \times 10^7$ and f successively 0.1, 1 and 10 to obtain $\delta = 207$ nm at 0.1 THz, $\delta = 66$ nm at 1 THz and $\delta = 21$ nm at

10 THz. All these penetration depths are much smaller than the vacuum wavelengths ($\lambda = c/f$), which are approximately 3 mm, 300 μm and 30 μm, respectively.

6.7 Summary

6.7.1 Key terms

dielectric function, 101
relative dielectric function, 101
conductivity, 102
driving force, 104
damping force, 104
restoring force, 105

conductor, 108
plasma, 110
refractive index, 112
complex refractive index, 114
absorption coefficient, 117
extinction coefficient, 117

6.7.2 Key equations

plasma frequency $\qquad \omega_p^2 \equiv \dfrac{nq^2}{\epsilon_0 m}$ (6.31)

effective dielectric function $\qquad \tilde{\epsilon} = \epsilon_0 \left[\dfrac{\omega_p^2}{\omega_0^2 - \omega^2 - i\omega\gamma} + 1 \right]$ (6.34)

effective conductivity $\qquad \tilde{\sigma} = \epsilon_0 \omega \left[\dfrac{\omega_p^2}{\omega\gamma + i(\omega_0^2 - \omega^2)} - i \right]$ (6.40)

refractive index $\qquad n(\omega) \equiv \dfrac{c}{v}$ (6.61)

relative dielectric function $\qquad \epsilon_r(\omega) = n^2(\omega)$ (6.64)

absorption coefficient $\qquad \alpha(\omega) = 2n_2(\omega)\dfrac{\omega}{c}$ (6.97)

6.8 Table of symbols, Chapter 6

General mathematical symbols appear in Appendix B. If the unit of a quantity depends on the context, this is denoted '—'.

Symbol	Meaning	Unit
A	general quantity	arbitrary
A_0	amplitude of oscillation	same as A
B	magnetic field	T
c	lightspeed	m Hz
E	electric field amplitude	V m^{-1}
f	frequency	Hz
f_0	frequency, fundamental	Hz
F	force	N
m	mass	kg
n	number of charges per unit volume	m^{-3}
n	(real) refractive index	[unitless]
$\tilde{n}$	complex refractive index	[unitless]
n_1, n_2	real, imaginary parts of complex refractive index	[unitless]
q, Q	charge	C
r	distance	m
R	reflectance	[unitless]
t	time	Hz^{-1}
v	speed	m Hz
z	distance	m
α	absorption coefficient	m^{-1}
δ	decay length, penetration depth	m
ϵ_0	electric constant	F m^{-1}
ϵ	dielectric function	F m^{-1}
ϵ_c	dielectric function, conductor	F m^{-1}
ϵ_r	relative dielectric function	[unitless]
μ_0	magnetic constant	N A^{-1}
σ	conductivity	S m^{-1}
σ_c	conductivity, conductor	S m^{-1}
ω	angular frequency	rad Hz
ω_0	angular frequency, fundamental	rad Hz
ω_p	plasma frequency	rad Hz

Part II

Components

7 SOURCES

This chapter employs trigonometry, differentiation, integration and vector algebra.

Electromagnetic radiation is produced

- by hot objects,
- by electric charges moving freely,
- by transitions between defined energy levels.

In Chapter 4, I discussed in detail the radiation given off by hot objects. Heat any object, and it will give off light. *Thermal sources* are perhaps the simplest and best-known sources of light. The sun radiates light because it is hot. Likewise the stars. The moon reflects the light of the sun. The embers of the fire glow because they are hot. Terahertz radiation is produced by thermal sources, but this will not be discussed in detail in this chapter; see Chapter 4 for further information.

An electromagnetic wave may be produced by 'waving', or appropriately moving, an electric charge, or charges. The principles behind the production of electromagnetic radiation by moving electric charges are set out in Section 7.2. The *synchrotron* and the *free electron laser* are two sources of terahertz-frequency electromagnetic radiation based on these principles.

Transitions between defined energy levels are the basis of the laser. The principles behind this are set out in Section 7.3. These principles inform the operation of the *Ge-laser*, the *quantum cascade laser* and the *molecular laser*, which all operate at terahertz frequencies. Moreover, many additional schemes for terahertz generation are based on using visible or near-infrared lasers, either operating in the continuous mode, or operating in the pulsed mode.

Thus Section 7.2 on *free charges* and Section 7.3 on *bound states* give the physical basis for the practical sources described in the rest of the chapter.

There are many, many sources of terahertz radiation. Section 7.1 gives an overview of terahertz sources, classified according to whether the principle of operation involves a thermal source, free charges in motion or lasers.

The remainder of the chapter gives more detailed information about ten terahertz sources.

Learning goals

Once you finish this chapter you should be able to describe terahertz sources

- based on vacuum electronics,
- based on bound states, that is, terahertz lasers,
- driven by lasers, both continuous and pulsed.

7.1 Survey of sources of terahertz-frequency electromagnetic radiation

There are many sources of terahertz radiation. Here I give a brief categorisation of the sources I will discuss in the remainder of this chapter.

7.1.1 Thermal sources – hot rods

Globar The main examples of thermal sources used in the terahertz regime are the globar, a silicon carbide rod heated to about 1500 K by passing an electric current through it, and the mercury lamp. The theory of thermal radiation has been given in Chapter 4.

7.1.2 Sources based on the motion of free charges – vacuum electronics

Synchrotron The synchrotron is a large particle accelerator. It typically takes up the area of a block or a football field, but may be larger. Charged particles are accelerated around a large circular path. Electromagnetic radiation is emitted tangentially. All frequencies are emitted, including terahertz frequencies.

Free electron laser Like the synchrotron, this is a large device, in which charged particles travel along an evacuated tube. The charged particles are electrons. The tube may be straight or curved. The action by which the terahertz radiation is emitted is a laser action. So the free electron laser bridges the categories of free carrier emitter and laser emitter.

7.1.3 Sources based on bound states – lasers

Germanium laser Although not widely used, this is a relatively simple type of laser based on a solid material. It directly produces terahertz radiation and in this respect is in the same category as the free electron laser.

Quantum cascade laser Like the germanium laser, the quantum cascade laser is a solid-state laser that directly produces terahertz radiation.

Molecular laser Like the quantum cascade laser, this directly produces terahertz radiation. In contrast to the previous lasers, this uses a gas as the operating medium. The laser uses another laser to pump it. This introduces the first of the indirect, or *pumped* terahertz emitters.

7.1.4 Sources driven by continuous lasers

Beat-frequency generation This is an indirect method of terahertz generation that uses two pump sources of slightly different frequency.

7.1.5 Sources driven by pulsed lasers

Photoconductive generation This is an indirect method of terahertz generation that uses a pulsed pump source to produce coherent terahertz pulses by a photoconductive mechanism.

Surface generation This is an indirect method of terahertz generation that uses a pulsed pump source to produce coherent terahertz pulses by a surface mechanism.

Photo-Dember generation This is an indirect method of terahertz generation that uses a pulsed pump source to produce coherent terahertz pulses by the differential mobility of electrons and holes in a semiconductor.

Optical rectification This is an indirect method of terahertz generation that uses a pulsed pump source to produce coherent terahertz pulses by an electro-optic mechanism.

7.2 Electromagnetic radiation from free charges – vacuum electronics

Free charges Not surprisingly, electric charges can produce electromagnetic radiation. We will see the conditions under which this can happen shortly. By 'free' charges, I mean electric charges that are not subject to a force (although this is not quite correct, as an acceleration is required, as will be explained soon). The charged particles that are usually involved are electrons, and I will concentrate on these, although other charged particles, such as protons, or ions, will follow the same principles. So when you think of a free charge, you might have in mind an electron travelling freely in space, not interacting with other matter. In describing an electron travelling freely in space, in other words, in a vacuum, it is common to use the term **vacuum electronics**. In practice, the electrons may be in an evacuated glass capsule, known as a *vacuum tube* or *valve*, or in an evacuated metal tube, in some sort of particle accelerator.

Accelerating charge An accelerating charge emits electromagnetic radiation. Not every charge emits electromagnetic radiation, only accelerating charges. For example, a charge that remains fixed in the same spot does not radiate. A stationary charge will give rise to an electric field – indeed, it gives rise to an electric field that permeates all space – but this is different from giving rise to electromagnetic radiation.

To emit electromagnetic radiation, an electric charge must change its velocity. Velocity is a vector; associated with a vector is not only a magnitude but also a direction; changing either the magnitude or the direction changes the vector. For example, if I am stopped at the traffic lights, and when they change, put my car into motion, and move first at 10 km/h, then at 20 km/h, then at 30 km/h, I am changing the size of my velocity and so am accelerating. As an example of changing the direction of my velocity, which is also called acceleration in mechanics, but may not be in common language, consider me driving around a roundabout at constant speed. I am first travelling north (at 30 km/h), then north-east (still at 30 km/h), then east (at 30 km/h). Since the direction of my velocity is changing, I am accelerating, even though the speedometer reads 30 km/h throughout. Of course, I may simultaneously be changing both the size and direction of my velocity, as when starting from a stop sign and turning a corner. In the following sections we will first consider acceleration related to a change of speed but while travelling in a straight line, then we will consider acceleration related to a change of direction while travelling at a steady speed.

Acceleration in the direction of motion We first consider a charge speeding up or slowing down as it travels along a straight line. We may refer to this as *linear acceleration*, meaning acceleration in a straight line. There are devices, called *linear accelerators*, built for this very purpose, of speeding up electrical charges, for example, for medical applications. A linear accelerator is essentially a long, straight tube that is evacuated and along which charged particles, such as electrons, travel.

I will now give the expressions for the electromagnetic radiation from a linearly accelerating electric charge. I will not derive these expressions, but simply state the results. The derivations are given in many electromagnetics textbooks. The key quantities that are involved are the charge on the charged particle, denoted q, and measured in Coulombs; the acceleration, a, in units of m s^{-2}, the physical constants of the speed of light in a vacuum, c, in m s^{-1}, and the electric constant ϵ_0, in F m^{-1} and the mathematical constant π. The power per unit solid angle, $dP/d\Omega$, is at a maximum in the direction perpendicular to the direction of motion. If you are unfamiliar with the concept of solid angle, you can just think of this as the power emitted in a cone pointing in a particular direction. Solid angle is measured in units of steradians (abbreviation: sr). There are 4π steradians in a full sphere or 2π steradians in a hemisphere. The power per unit solid angle perpendicular to the motion is

$$\frac{dP}{d\Omega} = \frac{q^2 a^2}{16\pi^2 \epsilon_0 c^3}. \tag{7.1}$$

The denominator in this equation is just a bunch of constants. The numerator shows that the radiated power varies quadratically with the charge, q. That is, if the charge doubles, the power radiated goes up by four times. There is a similar quadratic dependence on the acceleration, a.

Example 7.1 Electrons needed

How many electrons, accelerating at 10^{17} m s^{-2}, are needed to produce 1 nW of radiated power per unit solid angle at right angles to their motion? (These values for acceleration and power are in the ranges typically found in some terahertz sources.)

7.2 Electromagnetic radiation from free charges – vacuum electronics

Solution

To solve this problem, we apply Equation (7.1). Let us represent the required number of electrons by N. Each electron carries an elementary charge e. So the total charge $q = Ne$. Equation (7.1) then becomes

$$\frac{dP}{d\Omega} = \frac{(Ne)^2 a^2}{16\pi^2 \epsilon_0 c^3}.$$

We now rearrange the equation to make the unknown, N, the subject:

$$N = \sqrt{\frac{16\pi^2 \epsilon_0 c^3}{e^2 a^2} \left(\frac{dP}{d\Omega}\right)}.$$

We now substitute into the equation the known values of physical constants of $\epsilon_0 = 8.854187817 \times 10^{-12}$ F m^{-1}, $c = 299792458$ m^{-1} and $e = 1.602176487 \times 10^{-19}$ C; and the values given in the problem of $a = 10^{17}$ m s^{-2} and $dP/d\Omega = 10^{-9}$ W per unit solid angle. The resulting calculation gives, to three significant figures, N = 383,000.

Exercise 7.1 Radiated power

Ten million electrons are accelerated at $a = 10^{16}$ m s^{-2}. What is the radiated power per unit solid angle at right angles to their motion? ■

Exercise 7.2 The constant K

The physical and mathematical constants in Equation (7.1) may be collected up and the equation written as

$$\frac{dP}{d\Omega} = KN^2 a^2,$$

where N is the number of elementary charges. What is the value of the constant K in this equation? ■

The power radiated by a linearly accelerating electric charge is not the same in all directions. I have mentioned already that the power is maximum in the direction perpendicular to the travel. In the direction of travel, the power is zero, which may come as a surprise. If we introduce θ to represent the angle between the direction of travel and the point at which the power is measured, a more comprehensive expression for the power per unit solid angle is

$$\frac{dP}{d\Omega} = \frac{q^2 a^2 \sin^2 \theta}{16\pi^2 \epsilon_0 c^3}. \tag{7.2}$$

To reiterate the two key points I have just mentioned: in the direction of motion, there is no radiation (in that case $\theta = 0$ and so $\sin \theta = 0$); perpendicular to the motion, the radiation is greatest (in that case $\theta = 90°$ and so $\sin \theta = 1$).

Exercise 7.3 Rectangular and polar plots

Consider Equation (7.2). On a regular plot with $dP/d\Omega$ on the vertical axis and θ on the horizontal axis, plot $dP/d\Omega$ vs. θ. Now repeat, but using a polar plot; that is, with $dP/d\Omega$ being represented as the distance from the origin and θ being the angle from the forward direction. ■

The distribution of radiated power in different directions as calculated from Equation (7.2) is shown in Figure 7.1. The characteristic shape is generated by the $\sin^2$ term. This

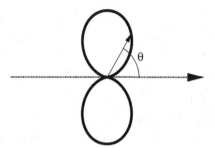

Figure 7.1 Electromagnetic radiation from a charge travelling in a straight line. This radiation pattern applies for a charge travelling at low speeds. Low speeds here means at speeds well below the speed of light in a vacuum.

may be pictured as a lobe, touching the charged particle, and rotated around the axis of travel. (Any object rotated around an external axis is called a *toroid* and so this is an example of a toroid. If the rotated object is a circle, the final figure can more precisely be called a *torus*. Loosely speaking it is a doughnut with a negligible hole. The charge moves along the doughnut axis.)

Example 7.2 Total power

Starting with Equation (7.2), determine the total power radiated.

Solution

To solve this problem, we need to add up the amounts of power radiated in different directions. This will require integration and also an understanding of the concept of solid angle. If you are unfamiliar with these, you will not be able to follow this example, but this is not necessary for understanding the material to follow. (Feel free to skip it.)

We obtain P by integrating over all solid angles:

$$P = \int_{4\pi} \frac{dP}{d\Omega} d\Omega .$$

Note, there is nothing specific here about P yet. This is the way any function is integrated over all angles. The integration is over a total of 4π steradians. We use the usual expression for the differential element of solid angle:

$$d\Omega = d\phi \sin\theta d\theta .$$

We combine the two previous equations to

$$P = \int_{\phi=0}^{2\pi} \int_{\theta=0}^{\pi} \frac{dP}{d\Omega} d\phi \sin\theta d\theta .$$

7.2 Electromagnetic radiation from free charges – vacuum electronics

The working so far would apply to any function P. At this point we introduce the specific expression we have from Equation (7.2) to give

$$P = \int_{\phi=0}^{2\pi} \int_{\theta=0}^{\pi} \frac{q^2 a^2 \sin^2 \theta}{16\pi^2 \epsilon_0 c^3} d\phi \sin\theta d\theta.$$

We now separate out the parts that do not vary with θ and those that do.

$$P = \frac{q^2 a^2}{16\pi^2 \epsilon_0 c^3} \int_{\phi=0}^{2\pi} d\phi \int_{\theta=0}^{\pi} \sin^3\theta d\theta.$$

The definite integral in ϕ is simply 2π. The definite integral in θ requires a little more effort to calculate, but turns out to be 4/3. Putting these values into the equation and simplifying,

$$P = \frac{q^2 a^2}{6\pi \epsilon_0 c^3}.$$

I haven't made a point of it up until now, but all the equations developed so far only apply to charges travelling slowly. By slowly, I mean at speeds much less than the speed of light. According to the special theory of relativity, we cannot accelerate anything past the speed of light. So the simple expressions I have given so far have to be modified to accommodate this fact.

In dealing with particle speeds relative to the speed of light it is conventional to introduce the quantity β, which is simply the ratio of the speed of the particle, v, relative to the speed of light, c. Since β is a ratio of two speeds, it has no units.

$$\beta \equiv \frac{v}{c}. \tag{7.3}$$

Exercise 7.4 β

Find the approximate value of β for (a) a sprinter moving at 9 m s^{-1}, (b) a plane flying at the speed of sound, 300 m s^{-1}, (c) a proton moving at half the speed of light. ■

In discussing relativity, it is also conventional to introduce the quantity γ, which is defined in terms of β. Thus γ, like β, depends only on the ratio of the speed of the particle to the speed of light. Like β, γ is dimensionless.

$$\gamma \equiv \frac{1}{\sqrt{1-\beta^2}}. \tag{7.4}$$

Exercise 7.5 γ

Find the approximate value of γ for (a) a sprinter moving at 9 m s^{-1}, (b) a plane flying at the speed of sound, 300 m s^{-1}, (c) a proton moving at half the speed of light. ■

Both β and γ increase as the object under consideration moves faster and faster. While β starts at 0 and moves towards 1 as a maximum, γ starts at 1 and increases without limit.

Now making our expression for the power generated per unit solid angle even more general by taking into account the special theory of relativity (I don't give the derivation here, only the final result):

$$\frac{dP}{d\Omega} = \frac{q^2 a^2 \sin^2 \theta}{16\pi^2 \epsilon_0 c^3 (1-\beta \cos \theta)^5}. \qquad (7.5)$$

In comparing Equation (7.5) with Equation (7.2) you will notice that the only difference is the introduction of the term $(1-\beta\cos\theta)^5$ in the denominator. If the charge is travelling slowly, then $v \ll c$, or, equivalently (from Equation 7.3), $\beta \sim 0$. Then Equation (7.5) will approach Equation (7.2), as we would expect. As the speed increases, and so β becomes larger, the side lobes of the radiation are thrown forward. (The radiation pattern is still technically a toroid.) This is illustrated in Figure 7.2.

Example 7.3 Optimum angle

Show that the angle at which the radiation is greatest is determined by

$$\cos\theta = (\sqrt{1+15\beta^2} - 1)/3\beta.$$

(Assume that $\beta \neq 0$.)

Solution

This derivation requires a knowledge of differentiation. If you are not familiar with differentiation, you will not be able to follow this example. This should not, however, prevent you reading on after this example.

We differentiate Equation (7.5) with respect to the variable θ to find the turning points. First, let's collect all the terms that do not depend on θ and write them as K'.

$$\frac{dP}{d\Omega} = K' \frac{\sin^2 \theta}{(1-\beta\cos\theta)^5}.$$

Now we differentiate this with respect to θ and set the result equal to zero. Considering the terms in θ to be a product of $\sin^2 \theta$ and $(1-\beta\cos\theta)^{-5}$ leads to

$$[2\sin\theta\cos\theta](1-\beta\cos\theta)^{-5} + \sin^2\theta[-5(1-\beta\cos\theta)^{-6}\beta\sin\theta] = 0.$$

We now multiply through by $(1-\beta\cos\theta)^6$ to obtain

$$\sin\theta[2\cos\theta(1-\beta\cos\theta) - 5\beta\sin^2\theta] = 0.$$

This equation has two solutions. The first solution is

$$\sin\theta = 0.$$

This solution corresponds to the minimum value of the power radiated; recall, there is no power radiated in the direction of motion. The second solution is

$$2\cos\theta(1-\beta\cos\theta) - 5\beta\sin^2\theta = 0.$$

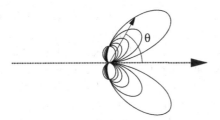

Figure 7.2 The speed of the charge, relative to the speed of light, is given by the quantity β. As the speed increases, the amount of radiation increases. As the speed increases, the angle of maximum radiation is thrown forward.

This solution corresponds to the maximum value of the power radiated. We may replace $\sin^2 \theta$ with $1 - \cos^2 \theta$ and then collect up terms to write the equation as

$$3\beta \cos^2 \theta + 2 \cos \theta - 5\beta = 0.$$

This is of the general quadratic form $ax^2 + bx + c = 0$ with $x = \cos \theta$. Applying the general quadratic formula gives the required result.

Exercise 7.6 Total power

(*Difficult.*) Starting with Equation (7.5), demonstrate that the total power radiated is:

$$P = \frac{q^2 a^2}{6\pi \epsilon_0 c^3} \frac{1}{(1-\beta^2)^3} = \frac{q^2 a^2}{6\pi \epsilon_0 c^3} \gamma^6.$$

Hint: To solve this problem, follow the procedure of Example 7.2. You will need to evaluate a definite integral of the form

$$\int_{\theta=0}^{\pi} \frac{\sin^3 \theta}{(1-\beta \cos \theta)^5} d\theta.$$

∎

Linear acceleration of a charge can produce electromagnetic radiation at terahertz frequencies. I will discuss practical ways for doing this with respect to beat-frequency generation (Section 7.6.1); photoconductive emitters (Section 7.7.1); and surface emitters (Section 7.7.2). In all these three types of practical emitters it is usual for the speeds of the charge carriers to be much less than the speed of light. So Equation (7.2) and Figure 7.1 are relevant, rather than Equation (7.5) and Figure 7.2. Two key points to remember from our discussion for acceleration in a straight line are:

- as the acceleration increases, the power radiated increases even more strongly;
- there is no radiation in the direction of travel.

Acceleration perpendicular to the direction of motion Up to now, we have been considering a charge moving in a straight line. Now we will consider a charge moving on a curve. If an object moves on a curve, it is changing the direction of its velocity. In other words, it is accelerating. (Remember, it is accelerating even if its speed is not changing.) To separate the acceleration due to changing speed from the acceleration due to changing direction, it is best to think about a situation where the charge is moving at a steady speed. The simplest curve is a circle. The curvature of the circle is the same everywhere. So, if you think about a charge moving at a steady speed around a circular path (uniform circular motion), you will have in mind a good testbed for this section. Here we are considering the radiation from a charge accelerating perpendicular to the direction of motion. This is exactly the situation in uniform circular motion. The charge moves in a direction along the circle, instantly tangential to the circle. The acceleration is directed to the centre of the circle, or radially. The acceleration is called *centripetal*, which means, literally, to the centre. So in uniform circular motion the acceleration is perpendicular to the direction of motion.

Just as there are linear accelerators, there are *circular accelerators*, although they do not usually go by that name. These are essentially an evacuated tube bent around to form a circle. Many particle accelerators are of this form.

I will now give, without derivation, the expressions for the power of the electromagnetic radiation from an electric charge in a uniform circular orbit. For simplicity, I will restrict the discussion to the radiation in the plane of the circle only. The analogous expression to Equation (7.2) is

$$\frac{dP}{d\Omega} = \frac{q^2 a^2 \cos^2 \theta}{16\pi^2 \epsilon_0 c^3}. \tag{7.6}$$

Exercise 7.7 Polar plot

Consider Equation (7.6). Plot $dP/d\Omega$ vs. θ on a polar plot, that is, with $dP/d\Omega$ being the distance from the origin and θ the angle from the forward direction. ∎

Equation (7.6) is exactly complementary to Equation (7.2). Whereas for the charge moving in a straight line, all the radiation was to the side, and none straight ahead, for the charge moving in a circle, all the radiation is in the instantaneous direction of motion, and none to the side. A good way to think of this is that in each case there is no radiation along the direction of acceleration; the radiation is directed perpendicular to this. The radiation from a charge accelerating perpendicular to its motion is shown in Figure 7.3.

Exercise 7.8 Total power

Follow the method of Example 7.2 to show that in the case of uniform circular motion the total power radiated is again

$$P = \frac{q^2 a^2}{6\pi \epsilon_0 c^3}.$$

7.2 Electromagnetic radiation from free charges – vacuum electronics

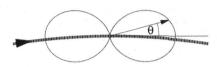

Figure 7.3 Electromagnetic radiation from a charge accelerating perpendicular to its motion. This radiation pattern applies for a charge travelling at low speeds. By low speeds is meant at speeds well below the speed of light in a vacuum.

Note: To solve this problem requires integration and an understanding of the concept of solid angle. If you are unfamiliar with these, you will not be able to complete this exercise, but this is not necessary for understanding the material to follow. ∎

You may have guessed already that Equation (7.6) only applies to charges travelling at low speeds, and you would be right. The full relativistic expression, analogous to Equation (7.5), is

$$\frac{dP}{d\Omega} = \frac{q^2 a^2}{16\pi^2 \epsilon_0 c^3} \frac{(\beta - \cos\theta)^2}{(1 - \beta\cos\theta)^5}. \tag{7.7}$$

The effect of increasing the speed of the charge, in other words, increasing β, is to increase the extent of forward-directed radiation at the expense of backward-directed radiation. This is illustrated in Figure 7.4.

Let us now concentrate on the power radiated specifically in the forward direction, in other words, for $\theta = 0$. In this case, Equation (7.7) becomes

$$\frac{dP}{d\Omega} = \frac{q^2 a^2}{16\pi^2 \epsilon_0 c^3} \frac{(\beta - 1)^2}{(1 - \beta)^5} = \frac{q^2 a^2}{16\pi^2 \epsilon_0 c^3} \frac{1}{(1 - \beta)^3}. \tag{7.8}$$

For a fast-moving charge, as β approaches 1, the forward radiated power increases rapidly.

Example 7.4 Forward power

How much power is radiated in the forward direction by an electron moving in a circle at half the speed of light, compared to an electron moving around the same circle at 75% the speed of light?

Solution
The two electrons differ only in their speeds. We will calculate β and γ for each of the electrons. For the first electron, from Equation (7.3), $\beta_1 = 1/2$. For the second electron, $\beta_2 = 3/4$. From Equation (7.8) we see the powers radiated will be in the ratio $(1 - \beta_1)^3$ to $(1 - \beta_2)^3$, or $(1/2)^3$ to $(1/4)^3$. The ratio is 2^3, or 8. Notice that the power has gone up eight times, even though the speed has only increased by 50%.

Figure 7.4 Electromagnetic radiation from a charge travelling in a circle (at several speeds). The speed of the charge, relative to the speed of light, is given by the quantity β.

Exercise 7.9 Forward power

How much power is radiated in the forward direction relative to the backward direction by an electron moving in a circle at half the speed of light?

Hint: Use Equation (7.7) with $\theta = 0$ for the forward direction and $\theta = 180°$ for the backward direction. ∎

The total power generated by a charge moving at arbitrary speed in uniform circular motion may be determined by following the method of Example 7.2. The result is:

$$P = \frac{q^2 a^2}{6\pi\epsilon_0 c^3} \frac{1}{(1-\beta^2)^2} = \frac{q^2 a^2}{6\pi\epsilon_0 c^3} \gamma^4. \tag{7.9}$$

Centripetal acceleration of a charge can produce terahertz radiation. I will discuss practical ways for doing this in Section 7.4.1 on synchrotron radiation. The key point from our discussion is

- in motion around a circle, there is radiation in the direction of travel and this increases greatly as the charge travels at higher speeds.

Exercise 7.10 Radiated power

(This exercise involves knowledge of vector algebra. It is not necessary to be able to understand or complete this exercise to continue beyond it.)

This exercise derives the specific equations that underpin the previous sections from the general equation for the power per unit solid angle radiated by a charge undergoing any sort of motion.

Let the velocity of the charge be $\mathbf{v}$ and the acceleration be $\dot{\mathbf{v}}$. (We are not making any assumption here about the relative directions of $\mathbf{v}$ and $\dot{\mathbf{v}}$.) Let the unit vector pointing from the charge to the point in space of interest be $\hat{\mathbf{r}}$. The power per unit solid angle at that point in space is

$$\frac{dP}{d\Omega} = \frac{q^2}{16\pi^2 \epsilon_0 c^3} \frac{[\hat{\mathbf{r}} \times ((\hat{\mathbf{r}} - \mathbf{v}/c) \times \dot{\mathbf{v}})]^2}{(1 - \hat{\mathbf{r}} \cdot \mathbf{v}/c)^5}. \tag{7.10}$$

(a) Show that in the special case of acceleration in a straight line, this equation leads to Equation (7.5). (b) Show that in the special case of uniform circular motion, this equation gives Equation (7.7). ∎

Dipole radiation Up until now we have been dealing with the electromagnetic radiation from a single electric charge. Now we'll consider the radiation from a pair of electric charges of opposite sign. A pair of electric charges of opposite sign is known as an electric dipole. Consider a positive charge $+q$ and a negative charge $-q$ separated by a distance d. The *dipole moment* is a vector pointing from the negative charge to the positive charge. The magnitude of the dipole moment is given by

$$p = qd. \tag{7.11}$$

Like a single point charge, a dipole will produce an electric field (in all of space). Again, like a single point charge, a static dipole will not emit electromagnetic radiation. A changing second derivative with respect to time, the mathematical way of expressing an acceleration, is necessary for electromagnetic waves to be emitted.

The electromagnetic radiation from a dipole can be rather complicated, especially close to the dipole itself, a region in space termed the *near field*. In the *far field*, that is, at distances many multiples of d away from the dipole, the radiated electric field is less complicated. Let us consider a point that is a distance r from the centre of the dipole along a line that makes the angle of θ with the dipole direction. Again, I will present the result without deriving it. The electric field component of the electromagnetic radiation has magnitude

$$E = \frac{1}{4\pi\epsilon_0} \frac{\sin\theta}{c^2 r} \frac{d^2 p}{dt^2}. \tag{7.12}$$

Let us unpack this equation. As mentioned above, for radiation to be emitted, a static dipole ($dp/dt = 0$) or even a steadily changing one ($dp/dt = $ constant) will not suffice; we need an *accelerating* dipole ($d^2p/dt^2 \neq 0$). The angular dependence is as $\sin\theta$. So, in a similar manner to the single charge moving in a straight line, there is no radiation along the axis of the dipole, and maximum radiation at right angles to this direction. Lastly, the electric field drops off with distance as $1/r$.

Although the dipole may appear more complicated than a single charge as a source of electromagnetic radiation, in some respects it is as simple or even simpler. For example, a simple motion to consider is the two charges bouncing in and out, as two masses joined by a spring. Rather than kicking a charge in a straight line, or swinging it around in a circle, we have two charges undergoing simple harmonic motion. In simple harmonic motion, the dipole moment varies with time according to

$$p(t) = qd \sin \omega t, \tag{7.13}$$

and so the radiated electric field will have the form

$$E = \frac{\omega^2}{4\pi\epsilon_0} \frac{\sin\theta}{c^2 r} p. \tag{7.14}$$

We might compare Equation (7.14) and Equation (7.12). The dependencies on θ and r are the same. In Equation (7.14), the term ω^2 in electric field, which goes to ω^4 in electric field intensity, severely reduces the dipole output at lower frequencies. This can produce difficulties in making practical sources of terahertz radiation.

A good picture to have in mind is a plunger moving up and down on the surface of a pool, radiating water waves perpendicular to its up and down motion. (An unhelpful picture would be a plunger moving back and forth horizontally in a pool, although that, too, produces waves. It might be helpful to recall that the electromagnetic wave is a transverse, not longitudinal, wave.)

The concept of **dipole radiation**, especially with the additional assumption of harmonic variation of the dipole moment, is the basis for the discussion of many types of terahertz emitters. These are discussed in Section 7.6.1 (beat-frequency generation), Section 7.7.1 (photoconductive emitters) and Section 7.7.2 (surface emitters). We will see that in the photoconductive emitter the direction of the dipole is convenient for utilising the emitted terahertz radiation, whereas on the surface field emitter it is not.

These results for the dipole hold only when the speed of the charges is considerably less than the speed of light; this is the situation in practical terahertz emitters.

7.3 Electromagnetic radiation from bound states – lasers

It is a general principle of quantum mechanics that, whereas a 'free' particle may have any energy whatsoever, a 'bound' particle is restricted to only certain allowed energy states. A celebrated and historically important example is the theory of the hydrogen atom as developed by Niels Bohr. Although a free electron may have any energy, Bohr postulated that an electron bound to a proton in the hydrogen atom could only have particular values of energy. These corresponded to certain allowed *states* of the electron. Electromagnetic radiation was emitted when the electron made the transition from one state to another.

The concept of light being emitted during a transition from one allowed state to another is central to the operation of the **laser**. I will discuss the laser now. First I will give a general description, then illustrate it with the particular example of the carbon dioxide laser.

Lasers are important in the generation of terahertz radiation for two reasons. First, there are some types of laser that directly emit in the terahertz region. These include the free electron laser, the germanium laser and the quantum cascade laser. Second, lasers are used to 'pump' other devices which in turn generate terahertz radiation. The great advances in terahertz science and technology in the last twenty years or so are centred on using very short pulses of light from pump lasers to produce synchronised pulses of terahertz radiation.

A solution looking for a problem The laser has been stereotyped as 'a solution looking for a problem'. Behind this epithet is the thought that the laser is an exquisite source of radiation – almost too good to be true – too good for workaday applications of light, needing an application that merits the laser's exquisite properties.

What is it about laser light that makes it so special? The main features are these:

- it is coherent – this means that the phase of the waves emanating from the laser retain their relative separation over a long time. In contrast, the waves from other common sources of light, such as the sun or from a fluorescent tube, are incoherent. Think of the crowd streaming from a railway station when the train pulls in; some leaving the train now, some later, jostling and bustling, compared to a drilled company of soldiers marching past a point. This illustrates the difference between an incoherent and coherent stream.

- it is monochromatic – this means it is of a single colour. Consider the company of soldiers, each soldier dressed in the same uniform, compared to the multicoloured, multistyled outfits of a normal street crowd.
- it is collimated – this means the light remains in a single beam or pencil rather than spreads as it travels. Compare a line of soldiers marching in a single file with a crowd dispersing in disparate directions.
- it is intense – this means a lot of light can be delivered to the same spot in a short time.

This unique combination of properties means lasers are widely employed across science and technology.

An instructive example: the carbon dioxide laser As an extended example to illustrate the laser principles in practice, I will now describe the carbon dioxide (CO_2) laser. CO_2 is a gas under normal conditions – we breathe it in, and breathe more of it out. So the CO_2 laser is an example of a gas laser. Just as there are gas lasers, there are lasers based on liquids and on solids. The working medium (in this case, CO_2) is termed the *lasant*. The CO_2 gas has states of different energies. There are various *vibrational* states. As the name implies, the states can be related to various to-and-fro oscillations of the carbon and oxygen atoms that make up the CO_2 molecule. There are two broad types of vibration: stretching, in which the atoms stay in the same straight line, and bending, in which the molecule is deformed from its original linear arrangement; in other words, when the O-C-O structure is no longer straight. Within the stretching motion, we distinguish symmetric stretching, when the two O atoms move in opposite directions (so, both away from or both towards the C atom) and antisymmetric stretching, when the two O atoms move in the same direction. Figure 7.5 gives the frequencies of the key vibrations of the CO_2 molecule.

When the molecule is not vibrating at all it is said to be in its ground state. This is indicated by the line at the bottom of Figure 7.5. Under suitable stimulation, the molecule may be made to vibrate in the antisymmetric stretching mode. The antisymmetric stretching mode has a frequency of 70.4 THz. We refer to this as an *excited* state of the molecule. In the context of our discussion here, this is the highest frequency vibration we will consider, so we may refer to it as the 'upper' state. It turns out that there are two other vibrational states between the ground state and the upper state. These are the symmetric stretching state, at 41.6 THz, and a bending state, at 38.43 THz. A molecule in the upper state may drop into these lower states by decreasing in frequency by (70.4 − 41.6 = 28.8) THz or by (70.4 − 38.4 = 31.9) THz. It needs to shed the excess energy, and may do so by emitting a photon, of frequency either 28.8 or 31.9 THz.

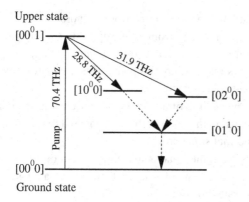

Figure 7.5 Key vibrations of the CO_2 molecule.

These photon frequencies correspond to photon wavelengths of about 10 μm and 9 μm, respectively. From either the symmetric stretching or the second bending state, the system may lose further energy to end up at the first bending state of 20.0 THz frequency; from there it will readily return to the ground state.

The lasing action depends on population inversion. Normally, the lower states have a greater population than the higher states. If a higher state is occupied more than a lower state, population inversion is said to have occurred. Population inversion occurs in the CO_2 laser because the upper state is rather long-lived relative to the lower states.

Rotations, as well as vibrations, play a role in the operation of the CO_2 laser. We refer to *rotational-vibrational* or *ro-vibrational* transitions. So far, we have only been considering the vibrations of the CO_2 molecule. If the molecule spins, then this introduces additional quantum states. The energy of the rotations is much less than the energy of the vibrational states. In this sense the rotations are a detail we have neglected so far in giving the 'big picture' of the operation of the CO_2 laser. Figure 7.6 shows the detail of the rotational structure added on top of the vibrational structure of the CO_2 molecule.

The rotational states are evenly spaced in frequency and labelled by a quantum number, usually denoted J. From details of quantum mechanics that I won't go into, the rotational states associated with the upper level take on only the odd values $J = 1, 3, 5 \ldots$; for the lower states, the values of J are even, $J = 0, 2, 4 \ldots$. Again, for reasons I won't go into, in changing from one rotational-vibrational state to another, the value of J must go up or down by one. So starting with $J = 1$, the molecule might finish up with $J = 0$ or $J = 2$. The transition to the higher frequency state is labelled 'P' and the transition to the lower frequency state is labelled 'R'. The J-value of the final state is given in parentheses at the end of the label. Transitions to the symmetric stretching state, of wavelength about 10 μm, are labelled '10'. Transitions to the bending state, of wavelength about 9 μm, are labelled '9'. For example, a transition starting at $J = 3$ and finishing at $J = 2$ in the symmetric stretching state is conventionally labelled '10R(2)'. Three other examples are given in Figure 7.6.

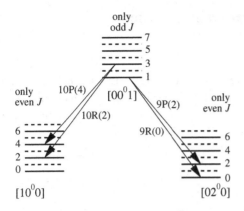

Figure 7.6 Rotational vibrations of the CO_2 molecule.

Example 7.5 Rotational-vibrational frequencies in the carbon dioxide molecule, CO_2

The rotational-vibrational transitions labelled 9R(20) and 9P(20) have frequencies 32.34 THz and 31.39 THz, respectively. From this data, what may be deduced about the rotational or vibrational frequencies in CO_2?

Solution

The rotational-vibrational transitions we have been discussing all start at the upper state of the antisymmetric stretching motion of frequency 70.4 THz. Those that are labelled beginning with 9 relate to a transition of photon wavelength about 9 μm to the second bending state of frequency 38.4 THz. The final number in the transition label, in this case 20, denotes the final rotational state of the molecule. So both the 9R(20) and 9P(20) transitions end at the $J = 20$ state. The two labels only differ in the letter, R or P. The letter relates to whether J is reduced by one in the transition (R) or increased by one in the transition (P). So the transition 9R(20) begins at $J = 21$ and the 9P(20) begins at $J = 19$. We thus deduce the separation of the rotational states associated with the antisymmetric stretching vibration to be (32.34–31.39) THz, or 0.95 THz.

Exercise 7.11 Rotational-vibrational transitions

The rotational-vibrational transitions labelled 9P(22) and 9P(20) have frequencies 31.33 THz and 31.39 THz, respectively. From this data, what may be deduced about the rotational or vibrational frequencies in CO_2? ■

Exercise 7.12 Rotational-vibrational transitions

The 10R(36) transition in CO_2 corresponds to a frequency of 29.54 THz. What is the frequency of the 9R(36) transition? ■

Exercise 7.13 Rotational-vibrational transitions

Consider the transitions labelled 10R(n), where n is an even number. Does the frequency increase or decrease as n increases? Explain. Do the same exercise for 10P(n). ■

7.4 Sources based on vacuum electronics

7.4.1 Synchrotron radiation

In a **synchrotron**, electrons travel in a large circular orbit at relatively uniform speed, but the constant change in heading leads to a beam of light being emitted. Synchrotrons can be made to produce very bright light, and all across the electromagnetic spectrum, from x-rays to microwaves.

7.4.2 Free electron laser

A distinctive type of laser is the **free electron laser** (FEL). It serves as something of a technological bridge between the *electronic* means of producing terahertz radiation and the *optical* means. By 'electronic' in this context, I mean 'vacuum electronics' – that is, manipulation of electric currents by massaging the flight of electrons. This technology was the origin of electronics, with the electrons being generated by thermionic emission, and then being sped up by electric fields provided by various electrodes, deflected by other electric and magnetic fields, and so on. To prevent the electrons being scattered by air, the electron path had to be in an evacuated vessel, the vacuum tube. Vacuum tubes, and vacuum electronics, dominated the first half century of electronics, and contributed amplifiers, rectifiers and logic circuits, among others. Now much of this has been supplanted by devices in which the electrons are confined in solid materials, typically semiconductors. The *free* in the FEL denotes that the electrons involved are travelling in a vacuum. No solid, liquid or gas lasing material is needed.

In fact, the electrons are not really 'free'. (By this I don't mean they are pricey, although it is true that free electron lasers are expensive.) If they were free, subject to no forces and so no acceleration, they would not radiate light.

In the FEL, electrons are wiggled. I'm serious. The electrons are accelerated, not by being suddenly slowed, as in an x-ray tube, or by being hurtled around a circle, as in a synchrotron, but by being wiggled from side to side. The means of wiggling the electrons is a series of magnetic fields pointing in alternating directions as shown in Figure 7.7. Such an arrangement of magnets is called a *wiggler* or an *undulatory*. The wiggler is usually made from strong permanent magnets.

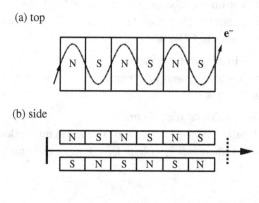

Figure 7.7 The heart of the free electron laser: the wiggler. The wiggler comprises an array of magnets. The north (N) and south (S) poles are indicated.

7.4 Sources based on vacuum electronics

Slaloming through the wiggler, the electrons radiate light. The direction of the radiation is along the wiggler, that is, in the overall direction of the electron travel. If it helps you, you might think of the electrons as travelling down through a series of mini-synchrotrons, first anticlockwise, then clockwise, irradiating always tangentially to their motion, and so on average producing a forward-pointing beam.

The frequency of light emitted by an FEL is determined by the electron speed and by the spacing of the magnetic field reversals. As before (Equation 7.3), $\beta = v/c$. We then define, as in Equation (7.4),

$$\gamma = \frac{1}{\sqrt{1-\beta^2}}. \qquad (7.15)$$

The term γ, like the term β, is a standard notation used in special relativity. Like β, γ involves only the particle speed v and the speed of light c. As the particle moves faster, γ increases.

The derivation of the frequency of light emitted involves relativistic calculation. I will simply state the result:

$$f = \frac{c\gamma^2\beta(1+\beta)}{w}. \qquad (7.16)$$

Here w is the spacing of the magnetic field reversals. This equation demonstrates that the frequency of the emitted radiation can be adjusted by varying either the magnet spacing w or the electron speed (changing β and γ). The FEL, then, has the desirable property of being continuously tuneable.

Exercise 7.14 Wiggler

Consider an FEL with electrons travelling at 90% of the speed of light ($v = 0.9c$). The magnetic field period spacing w is 5 cm. What is the frequency of radiation? To increase the frequency by 10%, what change would need to be made to (a) the electron speed? (b) the wiggler period? ∎

So much for the F and the E, what about the L? The discussion so far demonstrates how wiggling electrons radiate at terahertz frequency, but how is this a laser? A good question. We want to know how a *population inversion* occurs – how more electrons end up in the state from which they radiate light than from a random, thermal distribution. The short answer to this is that the light field in the wiggler cavity pumps the electrons into states where they are all in phase with the field. The light field in the wiggler (which could be supplied externally, say by a CO_2 laser, but need not be in practice) is transverse to the average forward direction of the electron motion, but in the direction of the electron motion as it travels over the 'synchrotron half loops'. In the half-loop motion, the electric field will either add to the electron motion or subtract from it. So the electrons are put into phase with the field. They bunch up. This bunched state is the *upper state* and the number of electrons in it represents the laser gain.

The FEL radiated power is related to the number of electrons. This follows from the previous paragraph. The electric field produced is directly proportional to the number of electrons and so the intensity is proportional to the square of the number of electrons. So doubling the electron beam current will increase the radiated power fourfold.

7.5 Laser sources

7.5.1 Germanium laser

Semiconductor lasers are based on energy states in semiconductors.

An early semiconductor laser was the heterostructure laser. In the heterostructure laser, both negatively charged particles (electrons) and positively charged particles (holes) are injected from one region to another region – hence the word *heterostructure*. The second region is termed the active region and it is here that population inversion occurs. The electrons and holes combine across the bandgap emitting light. Since both positively and negatively charged particles are involved, this laser is known as a bipolar laser. Another way of describing this laser is to say it involves transitions from the conduction band to the valence band and so is an interband laser. The problem in making such a laser operate in the terahertz region is to find a semiconductor with a sufficiently small bandgap.

Instead of looking to transitions between energy bands, transitions within energy bands may be used. The transitions within one band are known as intraband transitions. If the intraband transitions are in the conduction band, they only involve electrons. If the intraband transitions are in the valence band, they only involve holes. In either case, only one polarity of charge carrier is involved, and so the transitions are described as unipolar. Two examples of terahertz laser based on intraband, unipolar transitions are the germanium laser, under discussion here, and the QCL laser, to be described in Section 7.5.2.

The **germanium laser** is based on acceptor or p[ositive] doping and so for clarity and brevity will be referred to as the p-Ge laser. In the (original) p-Ge laser, crossed magnetic and electric fields are employed between light-hole and heavy-hole bands.

Another scheme involves cyclotron resonance transitions between Landau levels in the light-hole bands (specifically, from the $n = 2$ to $n = 1$ transitions of the b set of light holes). Depending on the doping, the tuning may be from 1 to 2 THz (light doping) or from 2 to 3 THz (heavy doping). Tuning is by varying the magnetic field on the Ge crystal. The need for low temperatures (4.2 K) and high magnetic fields (of the order of a few tesla) has meant that this type of laser has found limited application.

A third scheme involves resonant states induced by impurities in uniaxially stressed Ge. In a normal cystal of unstressed, undoped Ge, the heavy-hole and light-hole valence bands are degenerate at the zone centre. By applying a force in a particular crystallographic direction, the light- and heavy-hole bands separate – we say the degeneracy is lifted. Now, in adding an impurity, additional states are formed, such as the $1s$ state, close to the band edges. It is between these impurity states that lasing occurs.

7.5.2 Quantum cascade laser

Confinement of an electron in a semiconductor is possible by fabricating what is known as a 'quantum well'. The usual way to make a quantum well is to grow first one semiconductor material, then another, let's say, with a slightly smaller bandgap, then more of the original semiconductor. An electron thus sees a higher potential, then a lower, then a higher. The walls on the outside of the structure form the well; the confinement of the walls induces quantisation; hence the name, quantum well.

The quantum well structure provides a small additional quantisation to the electron which is already confined into a band of energies (the conduction band). The new, additional quantum states are referred to as minibands, in contrast to the existing and larger conduction band from which they derive. Transitions may occur between the minibands, just as they may occur between the conduction band and the valence band. It is these 'intraband' transitions, meaning within the conduction band, that are the origin of the radiation from a **quantum cascade laser** (QCL).

The QCL comprises a series of quantum wells. Electrons start at the first well, emit terahertz radiation, then are swept into the second well, where the process repeats. The electrons 'cascade' down through a series of steps, emitting terahertz radiation every step of the way.

The QCL is based on semiconductors, but there are three main ways in which it differs from a conventional semiconductor laser. First, the transitions are intraband rather than interband. This means they are of smaller energy and so better suited to terahertz operation. The operation is unipolar, only involving charged particles of one sign, in this case, electrons. Second, the spacing of the intraband energy levels may be varied by changing the dimensions of the quantum wells. Performing this in a controlled way is called *band structure engineering*. Whereas in a conventional semiconductor laser the frequency is determined by the bandgap, and to change the bandgap requires changing the material, in the QCL the quantum wells may be designed to provide the frequency of the radiation required. (In each case, though, the frequency is 'built in' to the device at design stage, and not easily changed after that.) Third, the conventional semiconductor laser has a one-shot operation – the electron drops from the conduction to valence band and is annihilated. In the QCL, the electron emits at the first well, then is used again at the second well, then at the third well, and so on.

A simple QCL has four levels. The electron is injected to the quantum well through tunnelling from level 4 to level 3. The injection is rather rapid. Once in the well, there is little choice for the electron to lose energy except by emitting a photon and dropping to level 2. One might say it is loath to do this; at any rate, the rate is rather slow. Then, from level 2, it is easily swept into level 1, then on out to the next injection area. By rapidly filling the wells from the top (levels 4 > 3), and rapidly sweeping out at the bottom (levels 2 > 1), a population inversion is built up between levels 2 and 3.

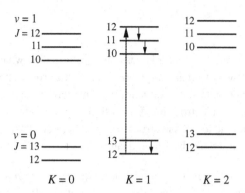

Figure 7.8 Lasing transitions in the optically pumped molecular laser.

7.5.3 Molecular laser

The optically pumped **molecular laser** operates in a similar way to the CO_2 laser, given as a prototypical example of a laser in Section 7.3. In the CO_2 laser, the energy to push the CO_2 molecule into the upper level is provided by an electrical discharge (and assisted by the presence of N_2 molecules). In the optically pumped molecular laser, the pumping is provided by the output of the CO_2 laser. This excites the molecule to a rotational-vibrational state. The molecule then loses energy through moving to a lower rotational state, emitting terahertz-frequency radiation as it does so.

Figure 7.8 gives one example of the lasing transitions in an optically pumped molecular laser. The example molecule is CH_3F, methyl fluoride. In fact, this was the first molecule to be demonstrated to produce a lasing transition in the terahertz regime on optical pumping. The lower part of the figure shows states in the lower vibrational state. The lower vibrational state is denoted by $v = 0$. As with the CO_2 molecule discussed earlier, there is finer structure to the allowed states related to rotation in addition to the vibration. As before, this may be denoted by a quantum number, J. (There is a further subtlety related to an additional quantum number K, but I will not discuss this.) Remember, as discussed in relation to the CO_2 laser, the spacing between the vibrational states is much larger than the spacing between the rotational states. For this particular molecule, the CO_2 pump line of 9P(20) is very close to the frequency difference between the $v = 0$, $J = 12$ and $v = 1$, $J = 12$ transition. So 9P(20) radiation will excite the molecule from the $v = 0$, $J = 12$ state to the $v = 1$, $J = 12$ state. In the $v = 1$ band there is now a greater population in the $J = 12$ state than in the $J = 11$ state. This population inversion will lead to lasing between the $J = 12$ and $J = 11$ state. (This, in turn, will lead to a population inversion between the $J = 11$ and the $J = 10$ state and subsequent lasing.) The separation between these states is of the order of 1 THz, and so a terahertz laser is produced.

There is another consequence of taking the molecule out of its original state, in this case, the $v = 0$, $J = 12$ state. This consequence is that this original state will be depleted in population relative to the next state of higher frequency, in this case, the $v = 0$, $J = 13$ state. So population inversion between these two states can result in laser emission. In this way a single pumping transition may lead to multiple terahertz lasing transitions.

The optically pumped molecular laser is tuneable in a limited sense. By changing the lasing gas, the gas pressure, the length of the laser cavity and the pump line, hundreds of different terahertz lasing transitions can be produced. These are, however, at the discrete specific wavelengths dictated by the particular molecule. The laser is not continuously tuneable, as, for example, a solid-state laser may be simply by adjusting the temperature of the laser crystal.

To obtain a satisfactorily high power, a gas tube of length 1 or 2 metres is used. The whole arrangement typically takes up a laboratory bench. Although you might find it simpler to think of the second laser tube following immediately the first laser tube, in practice, to make the whole arrangement more compact, the two tubes are placed side by side and mirrors used to steer the beam from the exit of the first laser to the entrance of the second. To introduce the gases and keep them at an appropriate low pressure, typically 1 mbar, gas piping, manifolds, valves and vacuum pumps are required. Water cooling is also required, and additional piping and pumping are needed for this.

In its favour, the optically pumped molecular laser shares many advantages common to any laser: the output is coherent, collimated and monochromatic. A distinguishing feature is that the output is intense. Commercial optically pumped molecular lasers typically boast several laser frequencies with powers of over 100 mW and perhaps a dozen or so with powers of over 10 mW. Research optically pumped molecular lasers have powers of over 1 W. This is an important advantage compared to other terahertz sources, which may only produce powers of a microwatt or nanowatt.

In concluding this section I will mention that it is possible to excite molecular lasers directly by an electrical discharge rather than by optical pumping. Direct excitation is simpler than optical pumping and so in the history of the field these lasers were developed before the optically pumped lasers I have discussed in this section. Of the many schemes for electrically excited molecular lasers, the most widely used were those based on the water (H_2O) and hydrogen cyanide (HCN) molecules. Optically pumped lasers have proved more powerful and versatile and so have largely superseded the H_2O and HCN lasers.

7.6 Sources driven by continuous lasers

7.6.1 Beat-frequency generation

The principle of **beat-frequency generation** is shown in Figure 7.9. Consider two harmonic waves of slightly different frequencies, f_1 and f_2, shown at the top of panel (a). For simplicity the waves have been shown with the same amplitude.

We have seen from Chapter 3 the result is a beating phenomenon described by Equation (3.29):

$$A_1 + A_2 = 2A_0 \cos[2\pi \frac{f_1 - f_2}{2} t] \cos[2\pi \frac{f_1 + f_2}{2} t + \delta]. \tag{7.17}$$

The original two frequencies yield two characteristic frequencies in the sum: an 'inner' frequency, the average of the original frequencies, $(f_1 + f_2)/2$, and an 'outer' frequency,

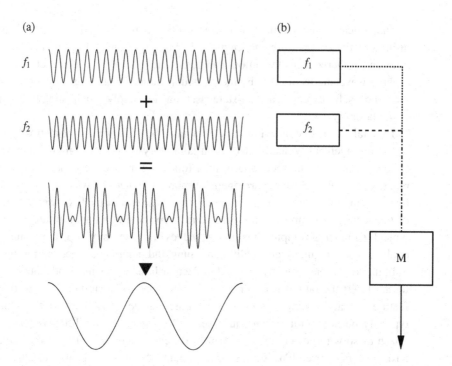

Figure 7.9 Generation of terahertz radiation using beats. Two frequencies, f_1 and f_2, are combined in a mixer (M).

half the difference of the two original frequencies, $(f_1 - f_2)/2$. If we are interested in the modulation of the outer envelope, or if we are interested in the electric field intensity, proportional to the electric field squared, then the final frequency of interest is simply the frequency difference $(f_1 - f_2)$.

The beat phenomenon may be used with light sources of two different frequencies to produce a lower frequency, as shown in Figure 7.9b. For example, two visible lasers, with frequencies of 350 and 352 THz, may be mixed to produce a beat frequency of 2 THz.

To generate a lower frequency by mixing two higher-frequency sources in this way is called difference-frequency generation (DFG) or beat-frequency generation (BFG). It is also referred to by the terms *heterodyning*, *optical heterodyning* or *optical heterodyning down conversion*. The term heterodyning has its origin in mixing two radio signals of high frequency to produce a lower frequency. Since two frequencies, or 'colours', are used, the method is also referred to as *two-colour mixing* (TCM).

Many different source lasers have been employed. In the early days, dye lasers were common. These were superseded by titanium:sapphire lasers. More recently, these are being displaced by laser diodes. It is usual that two separate lasers be used, but there are methods to extract two frequencies from a single laser. There also are methods to extract a multiple set of frequencies from a single laser. This allows many beat frequencies to be generated simultaneously.

Early systems used stand-alone lasers with the beams directed using conventional optical components such as mirrors and lenses. This is referred to as *free-space* coupling. As the technology has developed, it has become more common to couple the lasers using optical fibres. This has several advantages: the alignment of optical components is not critical, the lasers can be moved relative to one another, the laser beams are confined and so present less of a safety hazard. The system is more robust and stable. Now it has become possible to take the integration further and fabricate both lasers – and all the optics – on a single chip.

The device shown as M in Figure 7.9 serves to extract the terahertz-frequency envelope from the sum signal. The two principal ways to do this are to use a photomixer operating on a photoconductive emitter, the topic of Section 7.7.1, or to use a nonlinear crystal, the topic of Section 7.7.4.

7.7 Sources driven by pulsed lasers

7.7.1 Photoconductive emitters

The principles of photoconductivity are shown in Figure 7.10. We start with a piece of photoconductive material, Figure 7.10a. Typically a semiconductor, such as GaAs, is used.

Next, we form two electrical contacts on the photoconductor, Figure 7.10b. These electrodes are typically made of a metal, such as nickel or gold.

Then we set up a potential difference across the electrodes, Figure 7.10c. Typically, a cell or battery is used. A battery is characterised by the electrical potential difference

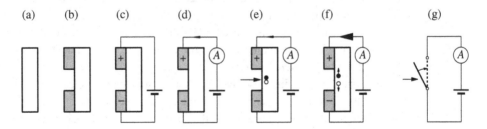

Figure 7.10 Photoconductivity principles. (a) A slab of photoconductive material. (b) Two electrodes are fabricated on the photoconductive slab. (c) A potential difference, V, is set up between the electrodes. (d) The current, I, flowing in the circuit may be measured by an ammeter. The conductance, C, is calculated from the current and the voltage by $C = I/V$. In a practical photoconductor, the current is small and so the conductance is small. (e) Light shines on the photoconductor. This creates charge carriers in the material. For example, an electron (shown as the full circle) and hole (shown as the empty circle) pair might be created. (f) The charge carriers increase the conductance. Negative charge carriers (for example, electrons) head to the positive electrode. Positive charge carriers (for example, holes) head to the negative electrode. Each contributes to the current. Now a large current flows for the same potential difference; the conductance has been increased. (g) Equivalent circuit. The light can be pictured as flicking a switch. When the light is off, the switch is open. When the light is on, the switch is closed.

between its two terminals. The potential difference is denoted V and is measured in volts. For an ideal battery the potential difference does not change, regardless of the load on the battery, and we will assume that to be the case here. We denote the higher potential electrode + and the lower potential electrode −. An alternative terminology is to say we are applying an electrical *bias*.

The potential difference across the photoconductor will cause an electric current to flow; the current may be measured by an ammeter, Figure 7.10d. We denote the current I. It is measured in amps. The ratio between the voltage and current is the resistance

$$R \equiv V/I, \tag{7.18}$$

measured in ohms. Of more interest is the inverse of the resistance, the conductance, measured in mhos,

$$C \equiv I/V. \tag{7.19}$$

Typically, the current is small, and so the conductance is also small.

Now we turn on the light, Figure 7.10e. The light produces electrically charged particles in the photoconductor. Typically, a pair of electrically charged particles is produced, a negatively charged electron and a positively charged hole, although this detail is not important to the principle of operation. (It is more accurate to say the electron and hole are made free or mobile by the light; they already are there in the material, but set free by the light.)

Moving charge is what makes an electric current. The charge carriers released by the light now move under the applied potential, Figure 7.10f. Negative charges will be attracted to the positive electrode, positive charges to the negative electrode. Conventionally, the motion of positive charge gives the direction of the current. A negative charge moving in the opposite direction gives the same direction of conventional current, and that is what we have here. A large current flows. The voltage has not changed, but the current has greatly increased and so (Equation 7.19) the conductance has increased greatly. The material becomes more conductive when light shines on it. This is why it is called a *photoconductor*.

The physical process just described can be represented by the electric circuit of Figure 7.10g. In the dark, the resistance is high, the conductance is low, it is as if the circuit is open. In the light, the resistance is low, the conductance is high. Light closed the switch. For this reason, an arrangement like this is called a *photoconductive switch*.

Photoconductivity can be used as the basis for detecting light. For example, photoconductive detectors are used in cameras to judge the illumination of the scene to be photographed. But how is photoconductivity used to produce light, specifically, terahertz radiation? The answer is, when a pulse of pump radiation hits the biased photoconductor, the charge carriers accelerate and radiate at terahertz frequencies.

Let's see how the **photoconductive emitter** works in Figure 7.11. Figure 7.11a shows the situation in the dark. The external circuit is not shown, but a potential difference is supplied across the electrodes. Figure 7.11b shows the arrival of the pump pulse and photocreation of charge carriers. The pump pulse is of short duration, of the order of a picosecond or less. Figure 7.11c shows the charge carriers spawned by the pump beam

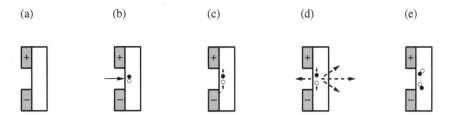

Figure 7.11 Role of charge carriers in photoconductive emitters. (a) Anticipation. (b) Creation. (c) Acceleration. (d) Radiation. (e) Recombination.

being accelerated in the electric field of the electrodes. Figure 7.11d shows the accelerating charge radiating electromagnetic radiation. The frequency of the radiation is in the terahertz range. Radiation is given off in many directions, but is preferentially coupled out through the high-index photoconductor. Figure 7.11e: very quickly electrons and holes recombine and we return to where we began in Figure 7.11a. There are many electron-hole pairs created and so many possibilities for recombining. Only one pair is shown in Figure 7.11(b–d), but two pairs are shown in Figure 7.11e.

Having established the general principles by which a pulsed pump beam can generate pulsed terahertz radiation, let's look into some of the practical details.

Candidate materials for photoconductivity are judged against several criteria.

- It is essential that the pump radiation can excite the conductivity by creating charge carriers. In the case of semiconductors, this means that the energy gap (bandgap) be less than the energy of the pump photons. Often titanium:sapphire lasers are used as the source of pump radiation. The light from the laser is 380 THz. There is a lot of interest is using other lasers, for example communications lasers that operate at around 190 THz (1.5 µm wavelength) for photoconductive emitters; a challenge is to find materials with a small enough bandgap.

- It is desirable for the material to have a large electrical resistance initially. In other words, it is desirable for the material to have a large dark resistance. This way, the change of conductivity will be most noticeable. Put differently, if the material is already very electrically conducting before the light strikes it, there will be little scope for the conductivity to change on illumination. Generally speaking, semiconductors may be made less conductive by removing sources of electrons (and holes). Foreign atoms in the semiconductors, called impurities, add electrons (or holes). Generally speaking, the dark resistance is increased by making the material purer.

Now let's move to the stage where the terahertz radiation is generated. Soon after the newly charged particles, electrons and holes, are created, they are swept towards the electrodes. This rapid motion of the charge carriers leads to a rapidly changing current. This changing current is the origin of the terahertz-frequency radiation.

The equation governing the emission of terahertz radiation is

$$E_{\text{THz}} \propto \frac{\partial J}{\partial t}. \tag{7.20}$$

For this current to produce electromagnetic radiation of terahertz frequencies, it is necessary that it change fast enough. For this to happen, the electrons and holes need to be able to move fast enough. This brings us to another property needed for a good photoconductive emitter.

- Electron and/or hole mobility must be high for a good photoconductive emitter. *Mobility* here is used in a technical sense to mean the speed of the charged particle relative to the electric field pushing it:

$$\mu = |v|/E. \tag{7.21}$$

The details of how much terahertz radiation is generated depend on many particular details of the particular experimental arrangement.

Finally, most pulsed systems are operated repetitively. That is, a pulse of light strikes the emitter, terahertz radiation is generated, then a short time later the process repeats itself. For example, a typical titanium: sapphire laser has a repetition rate (abbreviated as rep rate) of 80 MHz. This means a new pulse arrives every 12.5 ns. To operate under these conditions, the emitter material must be able to quickly return to its dark state. So the electrons and holes produced by the light in the previous pulse must be removed before the next pulse arrives.

- A short recombination time is desirable in a photoconductive emitter.

A larger bias will increase the terahertz emission, up to a point. The acceleration of the charge carriers is directly related to the magnitude of the electric field in the material which in turn is directly related to the voltage across the sample.

$$\frac{dv}{dt} = -\frac{v}{\tau} + \frac{q}{m}E. \tag{7.22}$$

If the bias is increased too much, however, the electric field becomes too high, and the semiconductor breaks down. A runaway occurs where a large current flows, damaging the material and reducing its resistance, allowing a larger current still to flow, and the damage to be exacerbated.

- A high breakdown field is a desirable property for a photoconductive emitter.

A similar breakdown can occur if the optical excitation power is too high. The deleterious effect of too high a bias and too high an optical pump power cannot be separated but must be looked at together. Each will lead to a greater current flowing in the photoconductor and so a greater joule heating of the sample. Heating a semiconductor will lead to a reduction in electrical resistance and so a further flow of current; positive feedback occurs, and irreparable damage may result. The thermal runaway can be averted or delayed in practice by cooling the emitter. For example, the emitter may be thermally attached to a block through which cooling water runs. The advantage of such an arrangement is that higher currents may be used and so higher terahertz powers emitted; the disadvantage is the additional complication to the experimental apparatus.

Many different designs for the electrodes have been tried. These include striplines, Hertzian dipoles and bow-tie geometries. Each has its own characteristics, yet almost any electrode pattern works, including painting dots of conductive paint on to the semiconductor surface. The details of electrode designs will not be discussed here.

We wish to couple the light out of the emitter efficiently. We will see in Chapter 8 that lenses can be used for this purpose.

7.7.2 Surface-field emitters

Terahertz radiation is emitted when charge carriers are accelerated in the surface field of a semiconductor. The operation of a **surface-field emitter** is similar to the mechanism of the photoconductive emitter we have just discussed. In both cases, a large population of charge carriers is produced in a very short time by the incoming laser pulse. In both cases, the photocarriers are accelerated by an electric field. In both cases, the rapidly changing current radiates an electromagnetic field at terahertz frequency. The difference between the two cases is the electric field. In the photoconductive emitter, the electric field is supplied by an external voltage source via electrodes fabricated on the emitter surface. In the surface-field emitter, the electric field is built in to the semiconductor surface itself.

Surface fields arise in semiconductors because the electronic structure of the surface differs from the electronic structure of the interior. There are many possibilities; I will describe a common one here. Let's start with the energy levels at the surface. The Fermi level at the surface is often located about halfway between the conduction and the valence band, as in Figure 7.12a. Consider an n-type semiconductor. In an n-type semiconductor the Fermi level is located closer to the conduction band than to the valence band. This is shown in Figure 7.12b. (In a p-type semiconductor, the Fermi level is located closer to the valence band.)

In Figures 7.12a and (b), to aid understanding, the surface and bulk are shown as separate, but in practice they are connected to each other. In bringing the surface and bulk states together, it is necessary that the Fermi level lines up. So the situation must be something like that shown in Figure 7.12c, where both the conduction band and the valence band increase in energy as the surface is approached. The terminology used in referring to this is *band bending*. It is said that the Fermi energy is *pinned* at the surface.

The result is an electric field pointing into the sample. To underline the orientation of the field: the surface electric field is perpendicular to the surface, in contrast to the photoconductive geometry, where the imposed electric field is parallel to the surface.

How big is the surface field? That depends. The potential difference will be determined by the details of the surface states, but let's assume they are mid-gap, and the doping of the semiconductor, but let's assume it is heavily doped, then it comes down to the semiconductor bandgap. GaAs has a bandgap of 1.4 eV, so we might expect a

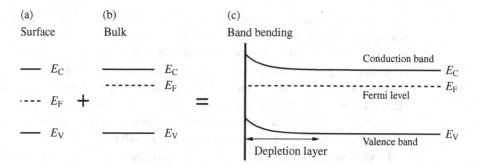

Figure 7.12 (a) Surface states in a semiconductor are usually found near the middle of the bandgap. (b) The Fermi level in an n-type semiconductor is found closer to the conduction band than to the valence band. (c) Since the Fermi level must be the same throughout, the bands bend near the surface. This phenomenon is also known as *Fermi level pinning*. The surface region in which the bending occurs and from which electrons are depleted is called the depletion layer. It is of the order of a micrometre in depth.

maximum surface potential of about 0.7 V. InAs has a much smaller bandgap of 0.4 eV, and so a much smaller maximum surface potential of about 0.2 V. Some *surface state engineering* can be done, but this does not usually result in a large gain. For example, doping GaAs layers with boron increases the potential by about 0.1 V. If we take the region in which band bending occurs to be 1 μm, then the maximum field is about 10^6 V/m, which is comparable to those used in photoconductive emitters.

The operation of a surface-field terahertz source is shown in Figure 7.13.

Let us now turn to some practical considerations. With respect to the depth of the surface field, it might be noted that the absorption depth of the typical pump radiation used (380 THz) in the typical semiconductors used (such as GaAs) is less that 1 μm – so all the photocarriers are produced in the region of the surface field.

You may have deduced from the last few paragraphs that the terahertz emission from the surface-field effect depends a lot on the surface field. From the point of view of

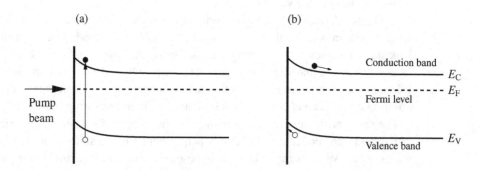

Figure 7.13 Surface-field emitter. (a) The pump beam promotes an electron from the valence band to the conduction band leaving a hole. (b) The electron slides down the conduction band slope into the bulk. The hole floats up the valence band slope towards the surface. A transient dipole results, radiating at terahertz frequencies.

making efficient emitters, a strong surface field in a material with the other desirable properties (high mobility, short recombination time) is wanted. But we may turn the situation around, and use the terahertz radiation emitted from the sample to tell us something about the surface field. For example, the polarity of the terahertz field will be reversed, going from a material with the surface field pointing in to one in which it points out.

A difficulty in the practical application of surface-field emitters is that the radiating dipole points in exactly the wrong direction! A practical way to overcome this, at least in part, is to arrange the geometry to collect terahertz radiation from other than the straight-through direction.

7.7.3 Photo-Dember emitters

The Dember effect refers to a dipole being created when electrons and holes separate due to their different rates of diffusion. The 'photo' part of the name refers to the origin of the electrons and holes, due to excitation by above-bandgap light. The electrons and holes are produced on the surface of a semiconductor. Usually, the electrons and holes diffuse at different rates. In most cases, the electrons diffuse much faster than the holes, so much so that the holes may be considered to be stationary. After some time, the 'centre of mass' of the electrons will have moved much farther from the surface than the holes. This will give rise to a dipole, which changes as the particles further diffuse. It is this rapidly changing dipole that gives rise to the terahertz-frequency radiation.

Radiation by a **photo-Dember emitter** is a surface effect. If the carriers are produced in the interior of a material, the centre of charge of both the electrons and holes is always at the point of creation. No dipole is formed. No radiation results.

Diffusion (of anything) is governed by the diffusion equation,

$$\frac{\partial N}{\partial t} = D \frac{\partial^2 N}{\partial z^2}. \tag{7.23}$$

We may write separate diffusive current equations for electrons and for holes.

$$J_e = -eD_e \frac{\partial \Delta N}{\partial z}. \tag{7.24}$$

$$J_h = +eD_h \frac{\partial \Delta N}{\partial z}. \tag{7.25}$$

(Remember, often the holes can be ignored, so this last equation may not be needed.) The diffusion constant is given by the Einstein relation

$$D = \frac{k_B T \mu}{e}. \tag{7.26}$$

Let's look at these terms in detail. There are several factors that will lead to a strong photo-Dember emitter.

- The ratio of electron to hole mobility. For example, this is much higher in InAs (30,000/240) than in GaAs (8600/400), not only because of the higher electron mobility, but also because of the lower hole mobility.

- The ratio of electron and hole temperatures. In narrow gap semiconductors, less of the incoming photon energy is needed to produce the electron-hole pairs, meaning more is left over to give them kinetic energy, or heat them. For example, in InAs, the electron temperature is about 8000 K, compared to about 800 K in GaAs. Because of the larger masses of holes than electrons, their temperature is less than the electrons. This lower temperature, as well as the lower mobility, contributes to the slower diffusion of holes relative to electrons.

In comparing photoconductive, surface-field and photo-Dember emitters, it may be seen that in each case an electric field appears: in photoconductivity, it is imposed externally; in the case of the surface field, it is due to Fermi level pinning; in the photo-Dember effect, it arises from charge carrier diffusion.

7.7.4 Optical rectification

Optical rectification strips the high-frequency or optical or inner structure from an optical pulse, leaving the low-frequency, terahertz, envelope.

Optical rectification is an analogous term to electrical rectification. If you are familiar with electrical rectification this may help you appreciate optical rectification. An electrical rectifier is used to convert an alternating current to a direct current. The rectifier essentially strips the high-frequency or alternating part from the incoming current and delivers a low-frequency (in this case, essentially zero-frequency) constant current.

Rectification is a **nonlinear** phenomenon. By this I mean the output of the rectifier does not bear a linear relationship to the input. A linear, or direct, or proportional output, denoted Y, to an input, denoted X, is written as

$$Y = a + bX. \qquad (7.27)$$

Many phenomena may be represented this way, especially if the input is small. But the relationship between the output and input may bear another, higher order term. The smallest such term is the quadratic term, the term in X squared:

$$Y = a + bX + cX^2. \qquad (7.28)$$

More generally, the output may contain terms of arbitrarily high order in X:

$$Y = a + bX + cX^2 + dX^3 + \ldots. \qquad (7.29)$$

but only the quadratic term is needed to introduce the concept of optical rectification.

Let's see how optical rectification works, first considering only a single frequency input, then we will look at two frequencies in the input. The single frequency input gives a close analogy with the electrical example, where an alternating input of 50 or 60 Hz, depending where you live, is converted to a direct, in other words, constant, current. We will assume that the input varies sinusoidally with time

$$X = E \cos \omega t. \qquad (7.30)$$

7.7 Sources driven by pulsed lasers

Let's not worry about the a and b terms here, but look directly at the first nonlinear term, the term in X^2.

$$X^2 = E \cos^2 \omega t. \tag{7.31}$$

It is a simple result in trigonometry (Appendix C) that $\cos^2$ may be written in terms of cos as follows:

$$E \cos^2 \omega t = \frac{1}{2} E(1 + \cos 2\omega t). \tag{7.32}$$

The two terms here correspond to a zero-frequency term and a frequency-doubled term. So, feeding in a quadratic dependence of a sinuosoidal term gives rise to a constant term and a term at double the frequency. If this were now to be run through a low-pass filter, only the constant term would remain.

Let's move this up to the next level of complication. Say two waves of frequencies ω_1 and ω_2 are combined. For simplicity, we will assume the two waves have the same amplitude. The linear terms will contain ω_1 and ω_2 only. The nonlinear terms will contain $\omega_1 \times \omega_1$, so just as we considered a moment ago, and $\omega_2 \times \omega_2$ – same. There will also be terms

$$X^2 = E \cos \omega_1 t \times E \cos \omega_2 t. \tag{7.33}$$

From basic trigonometry (Appendix C) this can be written as

$$\frac{1}{2} E^2 [\cos(\omega_1 - \omega_2)t + (\omega_1 + \omega_2)t]. \tag{7.34}$$

The first of these terms is called the difference-frequency term, the second, the sum-frequency term. So, combining two frequency sources in this nonlinear way can produce lower-frequency or higher-frequency terms. These are given the names difference-frequency generation (DFG) and sum-frequency generation (SFG). Sum-frequency generation is widely used in laser physics. Difference-frequency generation is a way to produce terahertz-frequency radiation from optical-frequency light.

Continuing in the same way, a spread of frequencies will give rise to a spread of difference frequencies and sum frequencies. This is the usual way with terahertz.

So, it is a matter of finding crystals with suitable nonlinear properties. The suitable crystals include inorganic and organic crystals. There are many interesting possibilities and a rich geometry depending both on the direction of the excitation beam, the direction where the radiation is detected, and the orientation of the crystal.

The method of optical rectification is qualitatively different from the other pulsed methods, photoconductivity, surface-field and photo-Dember. It is worth reviewing the differences. First, optical rectification is a nonresonant effect. Excitation across the bandgap is not needed. This makes it more widely applicable. Second, optical rectification does not require an external electric field. In this respect it is similar to the surface-field and photo-Dember mechanisms and different from photoconductivity. Third, particularly at low excitation levels, optical rectification is relatively inefficient. So a given amount of optical pump radiation is not converted into as much terahertz radiation. Fourth, as the excitation level increases, the efficiency of optical rectification relative to the other methods increases. This is for two reasons: the quadratic nature of the optical rectification means it goes up as E squared; while the others tend to saturate. Fifth, optical rectification can stand much higher pump power levels. So at very high power levels, optical rectification, while inefficient, will yield more terahertz radiation than photoconductivity in particular will, as the photoconductor will have been destroyed.

7.8 Summary

7.8.1 Key terms

vacuum electronics, 129
dipole radiation, 140
laser, 140
synchrotron, 144
free electron laser, 144
germanium laser, 146
quantum cascade laser, 147
molecular laser, 148
beat frequency, 149
photoconductive emitter, 152
surface-field emitter, 155
photo-Dember emitter, 157
optical rectification, 158
nonlinear, 158

7.8.2 Key equations

dipole radiation $$E = \frac{\omega^2}{4\pi\epsilon_0} \frac{\sin\theta}{c^2 r} p \qquad (7.14)$$

terahertz radiation $$E_{THz} \propto \frac{\partial J}{\partial t} \qquad (7.20)$$

7.9 Table of symbols, Chapter 7

General mathematical symbols appear in Appendix B. If the unit of a quantity depends on the context, this is denoted '—'.

Symbol	Meaning	Unit
a	acceleration	m Hz^2
a, b, c	constants	—
c	lightspeed	m Hz
C	conductance	S
d	separation of charges	m
D	diffusion constant	m^2 Hz
e	elementary charge	C
E_{THz}	terahertz electric field amplitude	V m^{-1}
f	frequency	Hz
f_1	frequency, first	Hz
f_2	frequency, second	Hz
K, K'	constants	—
k_{B}	Boltzmann constant	J K^{-1}
N	number of electrons	[unitless]
p	dipole moment	C m
P	power	W
q	charge	C
r	distance from dipole	m
R	resistance	Ω
t	time	Hz^{-1}
v	speed	m Hz
w	magnet spacing	m
β	speed ratio	[unitless]
γ	speed factor	[unitless]
ϵ_0	electric constant	F m^{-1}
θ	angle of radiation	rad
μ	mobility	m^2 V^{-1} Hz
ϕ	angle	rad
ω	angular frequency	rad Hz
Ω	solid angle	sr

8 OPTICS

This chapter is largely descriptive. It only makes use of rather simple mathematics: algebra and trigonometry (including the exponential function).

The last chapter was about sources of terahertz radiation. The next chapter will be about sensors of terahertz radiation. In this chapter we will look at what happens between the source and the sensor. We will look at how terahertz radiation is manipulated: bent, focussed, collimated. We will use *optics* as a general term for manipulating light and *optical elements* as a general term for specific devices that manipulate light.

Following the progression of an electromagnetic field through a collection of optical elements can be carried out at various levels of sophistication.

1. **Geometrical optics.** At the simplest level, light rays can be traced through the optical system using the approximation that they travel in straight lines and that the effects of wave properties, namely diffraction, are ignored. The assumption that wave properties are neglected means this is referred to as **geometrical optics** or *ray optics*. The assumption that the optical elements, in particular lenses, are thin, and that the angle of divergence of light from the optical axis is small – in other words, that only paraxial rays are considered – means this is referred to as *Gaussian optics*, in memory of the one who developed this method. Associated with Gaussian optics are fundamental geometrical formulas you might have met in previous studies that go by names such as the thin lens formula, the lens makers' formula, the mirror formula, the focal ratio or the f-number, formulas for magnifiers, microscopes and telescopes. Although conceptually and mathematically simple, this approach to optics goes a long way in accounting for the performance of even quite complex optical systems, and will serve in large part as the basis for the discussion in this chapter.
2. **Gaussian beam propagation.** In another and more sophisticated sense, the name Gauss is connected with the progress of light. This is when the light is not considered to be a thin or perhaps uniform cylinder or pencil of light, but rather to have a distribution of intensity perpendicular to the direction of propagation. In the simplest case the beam profile is Gaussian, in the sense that it falls off with an exponential dependence on the square of the distance away from the beam centre. More generally, the beam may have the profile of one or other of the Gauss-Hermite polynomials. *Gaussian beam propagation* is often a good representation of a laser beam. The emission of point-like terahertz

sources also often approximates this. Employing Gaussian beams, rather than geometric rays, is referred to as **quasi-optics**.
3. **Full calculation.** When the wave properties of light play a dominant role, there is little option but to carry out a full calculation of the electric field disposition through space by breaking up the volume into small elements and considering the evolution of the field over small time intervals. This approach is known as finite-difference time-domain analysis. It is typically carried out using a commercial software package. It is beyond the scope of this book.

Learning goals

After studying this chapter, you should be able to describe

- quasi-optics and their range of application,
- optical elements that redirect or reshape a terahertz beam: mirrors, lenses, lightpipes and prisms,
- optical elements that transmit, but may change the character of, a terahertz beam: attenuators, filters, windows and polarisers.

8.1 Quasi-optics

The profile of a quasi-optical beam is Gaussian in the transverse direction and hyperbolic in the propagation direction. Let's unpack this statement.

8.1.1 Transverse direction, transverse plane

The electric field of a Gaussian beam, as shown in Figure 8.1, varies with transverse distance x from the beam as

$$E = E_0 \exp\left[-\left(\frac{x}{w}\right)^2\right]. \tag{8.1}$$

The function $\exp(-ax^2)$ is named after Gauss. Here the constant a has been written as $1/w^2$ where w stands for the **beam width**. (Do not confuse the width, w, with the angular frequency, ω.) The function is symmetric about $x = 0$ and falls off to a value of $1/e$, or about a third, of the original at a distance of $x = w$. So it is more precise to refer to w as the half-beam width and call $2w$ the full-beam width. In fact, the function continues to have a finite value all the way to $x = \infty$, but for convenience we consider the width of the beam to be defined by w or $2w$.

The intensity is proportional to electric field squared and so we introduce a factor of two in the exponent in writing

$$I = I_0 \exp\left[-2\left(\frac{x}{w}\right)^2\right]. \tag{8.2}$$

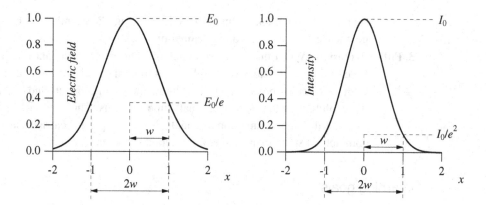

Figure 8.1 (a) The electric field of a Gaussian beam varies with transverse distance x from the beam as $E = E_0 \exp\left[-(x/w)^2\right]$. (b) The intensity of a Gaussian beam varies with transverse distance x from the beam as $I = I_0 \exp\left[-2(x/w)^2\right]$.

About 95% of the beam intensity is found between $x = -w$ and $x = w$.

We are assuming the beam is propagating in the z direction and so x is measured in the transverse direction. Equally, we could use the third perpendicular direction, y, in the equations. Or, measuring the distance from the beam axis z at some arbitrary direction in the transverse x-y plane, we could use $r = \sqrt{x^2 + y^2}$ instead.

8.1.2 Propagation direction, lateral plane

As the beam propagates in the z direction, as shown in Figure 8.2, the width varies as

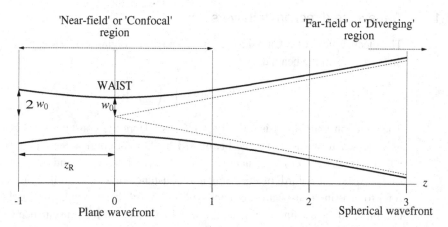

Figure 8.2 Gaussian beam. The minimum width occurs at the beam waist. Within z_R (the Rayleigh range) of the waist, the beam approximates a plane wave and the divergence is small; this is the *near-field* or *confocal* region. Far from the waist, $z \gg z_R$, the wavefronts become spherical and the beam diverges at a half angle that approaches w_0/z_R.

$$w(z) = w_0 \left[1 + \left(\frac{cz}{\pi w_0^2 f}\right)^2\right]^{1/2}. \qquad (8.3)$$

So the beam width depends on not only where we are along the propagation direction z, but the frequency f, as well as the minimum beam width w_0 – a constant for a given system – and the speed of light, c. We have set the origin of the z axis so that it corresponds to the minimum in width w; that is, $w(z = 0) = w(0) = w_0$. This minimum value of w corresponds to the narrowest part of the beam. The narrowest part of the beam is called the **beam waist**.

At the beam waist, the term in the square brackets is 1. How about when the term in the square brackets is 2? In this case,

$$w(z) = w_0 \sqrt{2}, \qquad (8.4)$$

which occurs for the value of z,

$$z_R = \frac{1}{2} k w_0^2 = \frac{\pi w_0^2}{c} f. \qquad (8.5)$$

This characteristic value of z is known as the **Rayleigh range**. It depends on π and on the beam waist width w_0 and, importantly, is directly proportional to frequency, increasing as f increases. The Rayleigh range defines the **near-field** or **confocal** region. Here the wave approximates a plane wave; its divergence is small.

Let us use the characteristic length in the propagation direction, z_R, and the characteristic length in the transverse direction, w_0, to simplify Equation (8.3). Dividing by w_0 and substituting for z_R we see that

$$\frac{w(z)}{w_0} = \left[1 + \left(\frac{z}{z_R}\right)^2\right]^{1/2}. \qquad (8.6)$$

Now introducing the dimensionless quantities $X = w(z)/w_0$ and $Z = z/z_R$ and rearranging, we have

$$X^2 - Z^2 = 1. \qquad (8.7)$$

This is the equation of a hyperbola. So the shape of the beam in the X-Z plane, in other words, in the plane containing the propagation direction and a direction perpendicular to that, is a hyperbola.

In the range of z closer to the beam waist than the Rayleigh range, in other words, in the range $-z_R < z < z_R$, the beam is relatively narrow, and may be considered paraxial. At relatively large distances from the waist, $z_R \ll z$, we may neglect the term of 1 in the square brackets and Equation (8.3) becomes

$$w(z) = \frac{cz}{\pi w_0 f}. \qquad (8.8)$$

Expressed as an angle of divergence we have for the divergence to one side of the line of propagation

$$\theta_0 = \frac{w(z)}{z} = \frac{c}{\pi w_0 f}.\qquad(8.9)$$

This may be called the half divergence; the full divergence is

$$2\theta_0 = \frac{2c}{\pi w_0 f}.\qquad(8.10)$$

This gives the divergence of the beam in the **far field**, well outside the Rayleigh range. For a given beam waist, the divergence decreases as frequency increases. Figure 8.3 shows the transition from confocal to divergent optics for two different half-widths.

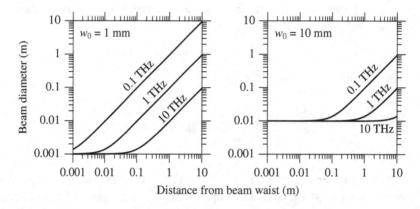

Figure 8.3 The transition from confocal to divergent optics for two different half-widths. For a beam half-width of 10 mm, the transition takes place approximately at a distance, measured in metres, corresponding to the frequency, measured in terahertz. This distance scales quadratically with half-width.

8.2 Mirrors

Perhaps **mirrors** are the simplest of all optical elements. Mirrors reflect light.

It is inbuilt into a mirror – one might say a consequence or demonstration of the law of reflection – that all frequencies of light are reflected at the same angle. We say there is no chromatic (colour) dispersion (separation). This feature is in sharp contrast to what happens in lenses. In lenses – one might say as a consequence of the law of refraction – different frequencies, in general, are refracted through different angles. Chromatic dispersion (or chromatic aberration) is a problem of lenses that mirrors do not have.

An ideal mirror reflects all the light that falls on it. This is saying two things – 100% of the light is reflected, and that is so no matter the frequency of the light.

There are many materials that reflect terahertz radiation. Metals generally are good reflectors of terahertz radiation. So, many mirrors for use in the terahertz regime are made from metal.

8.2 Mirrors

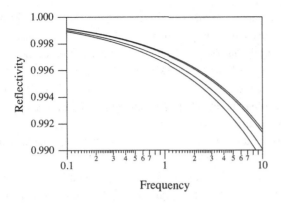

Figure 8.4 Reflectivity of highly conductive metals calculated from the Hagen-Rubens relationship. In order from the top, the metals are Ag, Au, Cu, Al. Below 8 THz, the reflectivity of all four metals is within 1% of perfect reflectivity.

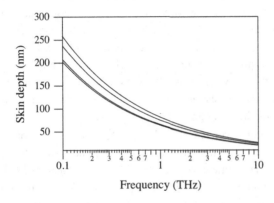

Figure 8.5 Skin depth or penetration depth of highly conductive metals calculated from the Hagen-Rubens relationship. In order from the top, the metals are Al, Cu, Au, Ag. In the frequency range shown, the skin depth of all four metals is less than 300 nm.

The reflectivity of a material increases with its conductivity. The best conductors are the best reflectors. The metals with the highest conductivity are, in decreasing order, Ag, Au, Cu, Al. Their conductivities are about 6.2, 5.9, 4.5, 3.8 $\times 10^7$ S m^{-1}, respectively. The reflectivity of these four metals is shown in Figure 8.4.

Of these metals, Ag and Cu tend to tarnish easily, and so they are used less often. Au, while being more reflective than Al, is rather more expensive.

The entire bulk of the mirror may be made from metal; alternatively, a layer of metal may be coated on another material, such as glass or plastic. Provided the metallic layer is thick enough, this will reflect as well as a bulk metal, and the hybrid mirror may well be lighter than the bulk mirror. A hybrid mirror may well be less expensive than a full metallic mirror; this is especially the case with Au. As shown in Figure 8.5, the skin depth for good conductors is less than 300 nm in the range 0.1 to 10 THz. So a metallic coating of a few micrometres thickness serves almost as well as a bulk material as far as reflection is concerned.

Apart from the advantages of high efficiency of throughput of incident light and of lack of chromatic aberration, mirrors have another property that makes them attractive and practical optical elements: they are relatively easy to manufacture and to use when made in large sizes. Large optical elements are useful in staying away from the physical optics regime and remaining in the geometrical optics regime. You may know that, while handheld telescopes often use lenses, large astronomical telescopes are usually based on mirrors instead. This is largely due to the mechanical consideration of the relative ease of manufacture and movement of large mirrors over large lenses.

As well as being good reflectors of terahertz-frequency radiation, these metals are also good reflectors of visible radiation. This is useful when aligning the optical system, as visible light can be used to trace the beam path. As we will see, lenses that are suitable for terahertz radiation do not always transmit visible radiation well, making alignment using visible light difficult or impossible.

The quality of a mirror surface is often stated by giving the deviation of the surface from the true in terms of the wavelength of light at which the mirror is intended to be used. A quality of $\lambda/20$ is often used. Since the wavelength of terahertz-frequency radiation is (much) longer than the wavelengths of visible-frequency radiation, the quality of the surface of a particular mirror is (much) better in the terahertz range than it is in the visible. For this reason, mirrors manufactured for use in the visible are often employed in the terahertz region. It is important to ensure, though, in the case of coated mirrors, that the thickness of the metal layer is sufficient for the planned frequency of operation (refer to Figure 8.5); a mirror designed for the visible range may have insufficient thickness of metal for the terahertz range.

One disadvantage of mirrors over lenses is that the central ray will continue in the same direction through a lens and this allows simple alignment of components in a *straight-through* geometry. The optical axis is a single line skewering the optical components, which may conveniently be serried on an optical rail. In contrast, a mirror deflects the central ray, so the optical axis changes direction. Alignment of the subsequent optical elements may be more difficult.

Three common shapes of mirror are the plane mirror, the parabolic mirror and the elliptical mirror.

A parabolic mirror takes rays parallel to the optical axis and concentrates them to a focus. Looked at the other way around, a parabolic mirror takes rays from a point source (at the focus) and sends these out as rays parallel to the optical axis.

The parabolic mirror therefore has two distinct uses. It may be used to focus a set of parallel rays and in this context is sometimes called a parabolic concentrator. Parabolic mirrors are used like this to collect solar energy. Conversely, a point source of light placed at the focus of the parabola will produce a collimated beam. Parabolic reflectors are used like this in such devices as torches, car headlights, spotlights and projectors.

In the applications of parabolic concentrators and parabolic reflectors just described, the parabolic surface used is usually symmetric around the optical axis. Another portion of the parabolic shape is widely used in terahertz optics, the off-axis parabolic mirror. This design simultaneously focusses and deflects incoming parallel rays (or, the other way around, simultaneously collimates and deflects the beam).

An ellipse has two focal points. An elliptical mirror takes rays from the first focal point and refocusses them at the second focal point. This arrangement is useful in conducting radiation from a point-like source to a point-like detector. As with the parabolic mirror, in practice an off-axis mirror is often used.

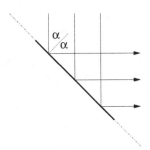

Figure 8.6 The plane mirror. The angle of reflection is the same as the angle of incidence (α). This is true for all frequencies of light; in other words, the mirror is not dispersive; in other words, the mirror has no chromatic aberration. Parallel beams striking a plane mirror remain parallel on leaving the mirror. If the angle of incidence is $\pi/4$, as shown here, the beam is turned through a right angle.

In three-dimensional space, the mirrors are generally surfaces and the rays do not lie in a single plane. Rotating the (two-dimensional) parabolic mirror about its optic axis resusults in a (three-dimensional) paraboloidal mirror; rotating the (two-dimensional) elliptical mirror around its optical axis results in a (three-dimensional) ellipsoidal mirror.

These are the main mirror types and each will be discussed in more detail in a moment. Of course, mirrors may be made in many other shapes; indeed, in arbitrary shapes. For example, circular and spherical mirrors have been constructed. These are rather simple both to make and to analyse, but they do not have the desirable property of taking rays from a point and either producing a collimated beam (as a parabolic mirror does) or refocussing the rays (as an elliptical mirror does). So we will not be discussing mirrors of spherical or more unusual shape. Instead, we will now look more closely at plane, parabolic and elliptical mirrors.

8.2.1 Plane

The plane mirror is shown in Figure 8.6. A **plane mirror** takes incoming parallel rays and deflects them into a different direction. The outgoing rays remain parallel to each other.

Two plane mirrors placed at right angles have the useful property that a light ray striking one of them is reflected back in the same direction that it came, as shown in Figure 8.7. This is provided that the light ray lies in the plane defined by the normals to the two plane mirrors (in Figure 8.7 this corresponds to the plane of the page). The returning ray will be displaced from the original ray, but the returning ray and the original ray will be directed along parallel lines. This useful property can be extended to three dimensions. Placing three mutually perpendicular plane mirrors together forms a *corner cube reflector*. You can picture it as the corner of a room, where the floor and two walls meet (presuming the walls are perpendicular to the floor and perpendicular to each other). Light striking a corner cube reflector from any direction is sent back in the opposite direction, but, in general, displaced from its original path.

Exercise 8.1 Lateral offset in the corner mirror

Calculate the offset of the returning ray from the incoming ray on a corner mirror as a function of the angle of incidence, α, and the distance from the corner, a, of the incoming ray.

Hint: See Figure 8.8. Determine d_a in terms of α and a. Then determine d_b. Finally, the lateral offset $d = d_a + d_b$. ∎

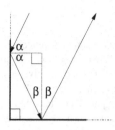

Figure 8.7 The corner mirror. Two mirrors are joined at a right angle. If the angle of incidence on the first mirror is α, the angle of incidence on the second mirror will be $\pi/2 - \alpha$, as seen from the right-angle triangle constructed using the two mirror normals. The result is that the outgoing beam is in the opposite direction to the incoming beam (and offset from it). The total deflection of the beam is through an angle of π. This holds regardless of the incident angle.

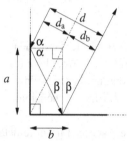

Figure 8.8 The corner mirror, showing the offset in the returning ray relative to the incoming ray.

Exercise 8.2 Corner cube reflector

Demonstrate that any light ray entering a corner cube reflector will exit the reflector in the opposite direction. ■

8.2.2 Parabolic

A **parabolic mirror** is shown in Figure 8.9. The only parameter needed to define a parabolic mirror is the parabolicity, a. Then we can write out, with z as the optical axis, and the base of the mirror at the origin, the equation of the parabolic surface as

$$z = a(x^2 + y^2). \tag{8.11}$$

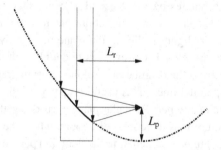

Figure 8.9 The parabolic mirror. Beams parallel to the optical axis striking the mirror all converge to the focal point. The focal point is a distance L_p from the base of the parabola. (L_p stands for parent focal length.) A parabolic mirror can be made from any part of the parabola. The section shown in bold here does not include the optical axis; this type of mirror is called an *off-axis* parabolic mirror. The central ray is deflected through a right angle. The effective focal length for the off-axis parabolic mirror is the distance from the centre of the mirror to the focal point. It is shown here as L_r, standing for reflected focal length. For any off-axis parabolic mirror, $L_r \geq L_p$. For the right-angle off-axis parabolic mirror, $L_r = 2 \times L_p$. The drawing is to scale. For example, with $L_p = 50.8$ mm, the whole vertical axis corresponds to 200 mm.

The many simple properties of the parabolic mirror follow directly from this equation. Parabolic mirrors are often used to collect light from a point source and convert it to a collimated beam, or vice versa, to collect a collimated beam and focus it to a point; Figure 8.10 illustrates these two uses.

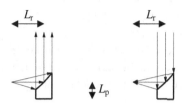

Figure 8.10 Parabolic mirror used to collect light from a point-like source and send it forth as a collimated beam and vice versa.

Exercise 8.3 Focal length of parabolic mirror

What is the focal length of the parabolic mirror specified by Equation (8.11)? ■

Exercise 8.4 Effective focal length of parabolic mirror

An off-axis paraboloidal mirror is formed from part of the surface given by Equation (8.11). What is the effective focal length (the distance between the mirror surface and the focus) for rays deflected through a right angle? ■

Coupling parabolic mirrors

Parabolic mirrors are often used in pairs, and often two pairs are used in a complete terahertz system. As we have seen, a single parabolic mirror can take light from a point source and collimate it. The first parabolic mirror in a terahertz system is often employed

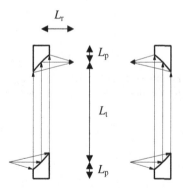

Figure 8.11 Two parabolic mirrors coupled together. The first mirror takes diverging light from a source and collimates it. The second mirror takes the collimated light and focusses it onto a sensor. Two arrangements are shown, zig-zag and folded. Each has the same efficiency.

in this role. The mirror takes some diverging light from a (point-like) source and sends the collected light all in the same direction. When the light is collimated in this way it is generally more useful than when it is heading in all directions; for example, it may be directed easily on to a sample, to measure the terahertz response of the sample. To collect the most radiation, a mirror with a large solid angle is required. In practice, with a mirror of reasonable size, this is achieved by having a short focal length. The off-axis section of the paraboloidal mirror is usually chosen so that the collimated beam has a circular cross section.

From the first parabolic mirror, the terahertz radiation may then traverse other optical elements or a sample before being detected. If the detector is point-like, a second parabolic mirror is often employed to focus the beam onto the detector.

Two possible geometries are shown in Figure 8.11. Optically, the two arrangements are equally efficient. The choice of which to use in particular circumstances comes down to convenience. The zig-zag configuration would be preferable if either the source or the detector needs room around it. The folded configuration is more compact.

Often it is required that the terahertz radiation be focussed at the sample rather than be in collimated form. In this circumstance, an intermediate focus is required between the source and the sensor. This may be achieved with an additional pair of parabolic mirrors, as shown in Figure 8.12.

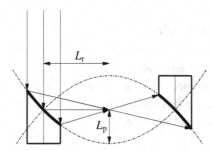

Figure 8.12 Producing an intermediate focus with a zig-zag configuration of parabolic mirrors.

Where the two off-axis paraboloidal mirrors form part of the same paraboloid, Figure 8.13, there is a slight inefficiency. Not all the light focussed from the first mirror is

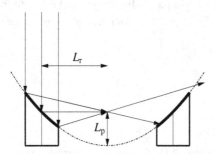

Figure 8.13 Producing an intermediate focus with a folded configuration of parabolic mirrors.

collected by the second. In the arrangement shown in Figure 8.12, all the light from the first mirror is collected by the second.

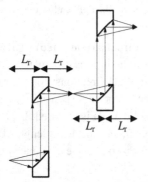

Figure 8.14 A convenient way of arranging four parabolic mirrors.

Often it is required to take the light from a source, focus it onto a sample, then collect it at a detector. Four parabolic mirrors are often used. These allow collection of radiation from a point-like source, refocussing at an intermediate position where a sample might be placed, then finally refocussing on a point-like detector. Many arrangements of two pairs of parabolic mirrors are possible. One is shown in Figure 8.14.

8.2.3 Elliptical

Two parameters are needed to define an ellipsoid: the semimajor axis length, a, and the semiminor axis length, b. Then we can write out, with z as the optical axis, and the centre of the mirror at the origin, the equation of the elliptical surface

$$\frac{(x^2+y^2)}{b^2} + \frac{z^2}{a^2} = 1 . \tag{8.12}$$

From this equation the fundamental properties of the **elliptical mirror** (Figure 8.15) may be derived.

Since a single ellipsoidal mirror can focus the rays from a point source to a point sensor, it plays the role of a pair of parabolic mirrors. The ellipsoidal mirror is not widely used in this role, however, as the beam is nowhere collimated.

An arrangement which allows radiation from a point source to be focussed, say at a sample position, then be re-collected and re-focussed at the sensor position, may be accomplished using two ellipsoidal mirrors. The second focus of the first ellipsoid is

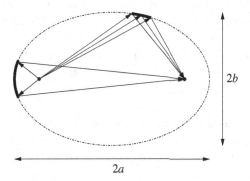

Figure 8.15 An elliptical mirror. The ellipse is characterised by the length of the major axis (the horizontal axis here), $2a$, and the length of the minor axis (the vertical axis here), $2b$. In general the ellipse has two focal points, indicated in the figure by the dots. Rays emitted from one focal point reflect to the other focal point. An elliptical mirror will therefore take rays from a point source at one focus and condense them at the second focus. This is illustrated by the mirror section to the left, an example of an axial elliptical mirror. A section of the ellipse may also be used to take rays from a point source and focus them while turning them through a specified angle. The section of the ellipse at the top illustrates turning through a right angle. Such a mirror is called an off-axis elliptical mirror. The right-angle off-axis mirror illustrated here is twice as far from one focal point as the other. In this case, the semimajor axis, a, is one and a half times the length of the semiminor axis, b.

made to correspond with the first focus of the second ellipsoid. It is at this position that the sample is set. The first focal point of the first mirror is the location of the source, the second focal point of the second mirror is where the sensor is placed. Such an arrangement is called a twinned or geminated ellipse. Although elegant in principle, this is not widely used in practice.

8.3 Lenses

As you may know, **lenses** are used widely in manipulating visible light, for example, in spectacles, telescopes and microscopes. The human eye itself contains a lens.

Lenses are not used so widely in the terahertz regime. There are several reasons for this. First, it is desirable that a lens transmit most (ideally, all) radiation falling on it. In the case of visible light, glass is a convenient lens material, with little absorption. There is not such a convenient material from which to make lenses in the terahertz. Although many plastics transmit terahertz radiation quite well, the small losses soon mount up after multiple lenses. Moreover, these plastics tend to have a small refractive index, requiring thicker lenses for reasonable focal lengths, compounding the problem of absorption losses. Second, if we consider high-refractive index materials, such as high-resistivity silicon, another problem arises: much of the incoming light is reflected from the front surface. In the case of visible optics, this reflection is overcome by anti-reflection coatings. Although anti-reflection coatings are available in the terahertz regime, they are not as highly developed as in the visible. Moreover, anti-reflection coatings normally only function efficiently over a limited range of frequencies. Although this is not much of a restriction over the visible span of an octave, 400–800 THz, it is much more difficult over the terahertz range of two decades, 0.1–10 THz. Third, lenses in general exhibit dispersion: that is, different frequencies of radiation are treated differently. The refractive index is frequency dependent. Another way to express the same thing is to say that lenses suffer chromatic aberration. Light of, say, 1 THz, will focus somewhere different from light of, say, 10 THz.

In spite of these difficulties, lenses find applications in terahertz optics. One advantage they have over mirrors is that the optical axis remains in a straight line before and after the lens. Alignment of the optical system is, literally, straightforward.

8.3.1 Lenses for focussing and collimating

Four lenses may be arranged to serve a similar purpose to the arrangements of mirrors in Figure 8.14. Light from the source is collected by the first lens. For a lens of given diameter, the largest solid angle will be attained by using the smallest focal length. The final lens serves the opposite role, focussing collimated light onto a point-like sensor. The second and third lenses provide an intermediate focus. These, in general, may have different focal lengths from lenses 1 and 4.

8.3.2 Lenses for refractive index matching

Total internal reflection

A consequence of Snell's law is total internal reflection. Consider a surface between two media of different optical density, such as silicon and air. Consider a ray originating in the more dense medium and striking the interface; in our example, a ray originating in the silicon. Let us consider Snell's law:

$$n_1 \sin \theta_1 = n_2 \sin \theta_2. \tag{8.13}$$

Say $n_2 > n_1$. To balance the equation, θ_1 must exceed θ_2. But there is a limit to how large θ_2 can become. The angle that the refracted beam can be bent from the surface normal can at most be a right angle. This condition defines the critical angle of incidence:

$$n_1 \sin \frac{\pi}{2} = n_2 \sin \theta_c, \tag{8.14}$$

or, rearranging to make θ_c the subject,

$$\theta_c = \arcsin \frac{n_1}{n_2}. \tag{8.15}$$

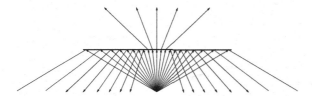

Figure 8.16 Total internal reflection. The upper part of the figure is a uniform medium of refractive index n_1. The lower part of the figure is a uniform medium of greater refractive index n_2. (The figure has been drawn for $n_2 = 3.5 \times n_1$, which corresponds to the ratio between the refractive indices of silicon and air at terahertz frequencies.) A source of rays is shown in the more dense medium. A ray that strikes the surface at normal incidence, that is, at zero angle of incidence, is undeflected. Rays that strike the surface at small angles are refracted away from the surface normal. The critical angle of incidence is given by $\theta_c = \arcsin(n_1/n_2)$; in this case it is about 0.29 radians or about 17°. Rays that strike the surface at an angle from the normal of greater than θ_c do not penetrate the surface but instead are reflected. This phenomenon is called total internal reflection.

Many terahertz sources are based on the production of terahertz radiation inside semiconductors. Many semiconductors have a refractive index between 3 and 4; let us take 3.5 as a representative value. From the previous equation, the critical angle can be estimated at $\theta_c = \arcsin(n_1/n_2) = \arcsin(1/3.5) \sim 0.29$ radians, or about 17° for a semiconductor in air. The high refractive index mismatch between the semiconductor and air means that there is a poor coupling efficiency between the inside of the semiconductor where the radiation is generated and the air outside where the radiation is required.

The inefficiency of coupling terahertz radiation from inside a semiconductor to the outside is illustrated in Figure 8.16. Only the rays from a narrow cone escape from the semiconductor surface. Rays that strike the internal surface from an angle of incidence exceeding θ_c suffer total internal reflection. Typically, only a few percent of all the rays produced by the source will escape the surface.

Exercise 8.5 Refractive index coupling

Assume a terahertz source located in a GaAs crystal emits radiation uniformly in all directions. Take the refractive index of GaAs to be 3.5 and the refractive index of air to be 1.0. What proportion of the radiation will be emitted through the surface of the crystal? ∎

The efficiency of emission from the surface may be improved by adding a convex lens of similar refractive index. For the lens, silicon is often used. If it is prepared to high purity, it transmits terahertz radiation very well. Silicon with few impurities has a low electrical conductivity, or a high resistivity; for this reason high-resistivity silicon is employed for terahertz optical components. A simple arrangement is half a sphere, as shown in Figure 8.17. Rays from beyond the original cone strike the final surface, now the outside of the silicon lens, at a lower angle of incidence to the angle they strike the original semiconductor surface, and are transmitted.

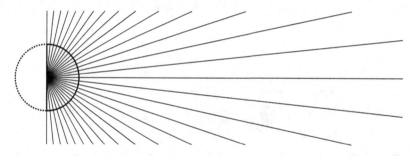

Figure 8.17 Hemispherical lens.

If the source is at the centre of the base of a hemisphere, the rays striking the spherical surface all do so at normal incidence and so suffer no refraction; they are undeflected. This has a considerable advantage over the flat surface in that all the rays emitted in the forward direction make it out of the semiconductor into air; none suffer total internal reflection. The rays are not, however, at all condensed or focussed by this arrangement.

The divergent rays emerging from the hemispherical geometry can be made to converge somewhat by moving the source a little away from the centre of the sphere. Two arrangements have special features and we will look at these.

Aplanatic lens An **aplanatic lens** is shown in Figure 8.18. The position of the source has been moved away from the centre of the sphere (to the left in the figures). Moving the source this way can be imagined to 'pull' the rays back, causing them to come close outside the sphere. The source has been moved to a special position that has a couple of unique features. First, the source has been moved back as far as possible before any rays suffer total internal reflection. That is, a ray moving at right angles to the optic axis within the sphere strikes the surface at the critical incidence angle and so is refracted along the surface. Second, the rays all apparently emerge from the same point; all the rays, when continued (imagining the sphere not there) would converge to a single point. In this sense, there is no aberration.

Collimating lens A **collimating lens** is shown in Figure 8.19. The position of the source has been moved away from the centre of the sphere and even farther than for the aplanatic lens. In doing so, the convergence has been made tighter. It has been moved to

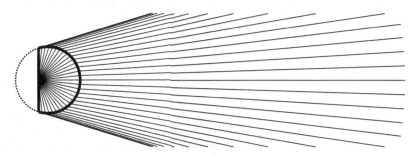

Figure 8.18 Aplanatic lens.

Figure 8.19 Collimating lens.

the position of the focus calculated in the paraxial approximation. You can see in Figure 8.19 that rays emanating from the source at a given angle end up close to the axis; that is, at a closer angle. The smaller the angle from the axis to begin with (inside the sphere), the closer to the axis the rays emerge (outside the sphere). At angles far from the axis, aberration, in this case *spherical aberration*, is evident; all the rays do not focus to the same point (ideally, at infinity), but focus closer to the sphere the greater the angle they began at. You may also notice that the high-angle rays do not escape the lens. They suffer total internal reflection (not shown in the figure). In this respect, the collimating lens is inferior to the hemispherical or the aplanatic.

Mathematically describing the hemispherical, aplanatic and collimating geometries is made with reference to Figure 8.20. For a sphere of radius R and source offset H, the total distance base to tip is $T = H + R$.

For the hemispherical lens, $H = 0$ and so $T = R$. All rays emitted from the source are normal to the surface and so no refraction occurs, regardless of the values of n_1 and

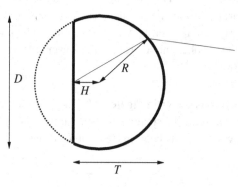

Figure 8.20 Lenses based on the sphere. The sphere has radius R and diameter D. The source is a distance H from the centre of the sphere; the line joining the source and the centre of the sphere defines the optic axis. The sphere is cut along the plane perpendicular to the optic axis through the source. The total length of the truncated sphere from base to tip is $T = H + R$. The diagram is drawn for refractive indices inside and outside the sphere in the ratio 3.5.

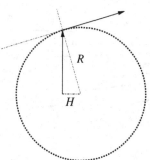

Figure 8.21 Aplanatic spherical lens. The offset H is determined by the condition that a ray leaving the source perpendicular to the optic axis is refracted through the maximum amount, namely, a right angle. The critical incidence angle satisfies the relation $\sin\theta_c = H/R$. Combining this with Snell's law, $n_1 \sin(\pi/2) = n_2 \sin\theta_c$, yields $H = Rn_1/n_2$. The diagram is drawn for refractive index ratio $n_2/n_1 = 3.5$.

n_2, regardless of the frequency or wavelength. There is no spherical aberration; outside the lens, all rays appear to diverge from the same point, which is indeed their point of origin. Apart from slowing the rays in their journey, the lens may as well not be there.

For an aplanatic lens, H is determined by the condition that a ray leaving the source perpendicular to the optic axis is just at the critical angle, and so is refracted through a right angle. This is shown in Figure 8.21. From Snell's law we have $n_1 \sin(\pi/2) = n_2 \sin\theta_2$; but the sine of the incidence angle is H/R, hence $H = Rn_1/n_2$. All the rays appear to diverge from a single point, located a distance $J = Rn_2/n_1$ from the centre of the sphere (to the left in the diagrams). The greatest angle the rays make with the optic axis is given by the critical angle $\sin\theta_c = n_1/n_2$. For silicon in air ($n_2 = 3.5$, $n_1 = 1$), this angle is abouts 0.29 radians or about 17°. This is the half cone angle; the cone angle is about 0.58 radians or about 33°.

For the collimating lens, H is determined by using paraxial optics to calculate the image position for an object at infinity ($n_1/\infty + n_2/s_i = (n_2 - n_1)/R$). The result is $H = Rn_1/(n_2 - n_1)$.

Some key features of the hemispherical, aplanatic and collimating lenses are tabulated in Table 8.1.

Example 8.1 Silicon lenses

Three high-resistivity silicon balls, each 10 mm in diameter, are available to make substrate lenses to couple to a GaAs photoconductor source. Assume, for simplicity,

8.3 Lenses

Table 8.1 Lenses based on slicing a sphere

Lens	Hemispherical	Aplanatic	Collimating
Offset, H	$H = 0$	$H = R\left(\dfrac{n_1}{n_2}\right)$	$H = R\left(\dfrac{n_1}{n_2 - n_1}\right)$
Thickness, $T = H + R$	$T = R$	$T = R\left(\dfrac{n_1 + n_2}{n_2}\right)$	$T = R\left(\dfrac{n_2}{n_2 - n_1}\right)$
Spherical aberration?	no	no	yes
Ray pattern	isotropic	within cone of angle arcsin (n_1/n_2)	approximately parallel to axis
Aplanatic focus		$J = R\left(\dfrac{n_2}{n_1}\right)$	

that both Si and GaAs have the same refractive index of 3.5 and the arrangement is to be used in air, $n = 1.0$. If the source is on the far side of a GaAs wafer 300 µm thick, where should the balls be cut so that the wafer on the truncated spheres yields a hemispherical, aplanatic and collimating lens?

Solution

We use the equations in Table 8.1 with the data $R = 5$ mm, $n_1 = 1$ and $n_2 = 3.5$.

This yields the results given in Table 8.2. To account for the thickness of the substrate, we would take 0.3 mm off the thickness of each lens, making them 4.7 mm (hemispherical), 6.129 mm (aplanatic) and 6.7 mm (collimating).

Table 8.2 Lenses based on slicing a sphere – results

Lens	Hemispherical	Aplanatic	Collimating
Offset, H	$H = 0$	$H = (5\,\text{mm})\left(\dfrac{1.0}{3.5}\right)$ = 1.429 mm	$H = (5\,\text{mm})\left(\dfrac{1.0}{3.5 - 1.0}\right)$ = 2 mm
Thickness, $T = H + (5\,\text{mm})$	$T = 5$ mm	$T = 6.429$ mm	$T = 7$ mm

Spheres are easy to manufacture to high precision, which in part accounts for the wide use of spherical objects. Other related lenses are the *bullet* lens, a hemisphere on top of a cylinder of the same radius. By suitable choice of the cylinder length, the bullet lens can be made to approximate the *elliptical* lens. The elliptical lens shares some of the elegant optical properties of the elliptical mirror, but is more difficult to fabricate than spherical objects. The bullet lens and the elliptical lens are examples of the class of *aspheric* lenses. These may be designed, typically using specialised computer code, for special purposes.

The ray-tracing method neglects diffraction and interference effects that are often important. We have also ignored reflections from the back of the lens. More sophisticated approaches to the radiation pattern from the lens take these into account, usually using the Fresnel-Kirchhoff diffraction integral. In the most complete treatment, an arbitrary radiation pattern (for example, from a dipole) and an arbitrary lens shape (for example, ellipsoidal) can be calculated numerically.

A final note on nomenclature. Mathematically, if a sphere is sliced through the middle, two hemispheres result. If a sphere is sliced anywhere but through the centre the smaller portion is called a hypohemisphere and the large portion a hyperhemisphere. In this sense, both the aplanatic lens and the collimating lens are hyperhemispheres. Some use the name hyperhemisphere as a shorthand for aplanatic hyperhemisphere, but this is best avoided, as it may lead to confusion with the collimating hyperhemisphere.

Exercise 8.6 Refractive index coupling

Estimate the additional amount of radiation emitted using a hemispherical lens relative to that emitted from the original surface. ∎

Exercise 8.7 Refractive index coupling

Comment on the angular distribution of the rays from the original surface and from the hemispherical lens, assuming the angular distribution at the source is isotropic. ∎

8.4 Lightpipes and waveguides

You may know that in the visible region, much information is transmitted up hill and down dale in optical fibres. These are typically made of a glass or plastic and are compact and flexible. No exact analogue exists in the terahertz regime.

One simple way to transmit terahertz radiation in a confined space is to use a circular pipe, even a copper pipe such as used in transferring water. This might be regarded as an optical or quasi-optical approach.

Another method, this time deriving from the low-frequency rather than high-frequency side of the terahertz regime, is to use waveguides, similar to those utilised with microwaves, but on a physically larger scale. Waveguides have been constructed in many geometries, for example, as parallel plates, rectangular cross-section pipes, and circular cross-section pipes.

8.5 Prisms

Prisms are widely used in visible optics. You may have heard of a prismatic compass. Binoculars usually contain prisms (double Porro prisms, often), which give their distinctive shape.

Prisms are used less widely in the terahertz regime, for much the same reasons as lenses: lossless and, preferably, dispersionless materials are difficult to find, key requirements for transmissive optical elements. Nevertheless, prisms find some terahertz applications.

An example of a prism is the right-angle prism with equal angles at the other two corners, Figure 8.22. A terahertz beam is shown entering from the left. It strikes the first

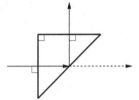

Figure 8.22 A right-angle prism.

face of the prism at zero angle of incidence. There is no refraction, regardless of the refractive indices inside and outside the prism; the beam proceeds in a straight line. (If there is a refractive index mismatch, some of the beam will be reflected back along itself, but we ignore that here.) When the beam strikes the hypotenuse of the prism, in general, some light will be reflected. For the geometry shown here, the angle of incidence at the hypotenuse is half a right angle and the angle of reflection is the same; the total deflection of the beam is through a right angle and it strikes the exit surface of the prism at normal incidence so, as with the entrance surface, proceeds through undeflected.

If the angle of incidence at the hypotenuse is large enough, the beam will undergo total internal reflection. How large is large enough? The requirement follows from Snell's law, that

$$n_2 \sin(\pi/4) > n_1 \sin(\pi/2), \qquad (8.16)$$

where the left-hand side of the equation refers to inside the prism and the right-hand to outside. This simplifies to:

$$n_2 > n_1 \sqrt{2}. \qquad (8.17)$$

Taking $n_1 = 1$ for air, this requirement is then that the prism material have refractive index greater than ~1.4, which is easy to realise in practice. A typical prism material is silicon, for which the refractive index is 3.5, for example. So the prism, without any metallic coating or other modification, can serve as a corner reflector, although, as mentioned before, reflection and (small) absorption losses mean this is not common.

Evanescent waves

What about the dotted beam shown piercing the hypotenuse in Figure 8.22? If the condition for total internal reflection is not met – that is, if the prism index of refraction is not too much greater than the index of refraction of the surrounding area – some light will penetrate the boundary in this direction. The interesting thing is, even if the condition for total internal reflection is met, some radiation does penetrate through this boundary, although it decays away rapidly. This escaping radiation is termed an **evanescent** (meaning, vanishing like a vapour) wave.

I won't derive here the expression for the penetration depth of the evanescent wave, but simply state it:

$$d = \frac{c}{2\pi f n_2 \sqrt{\sin^2 \theta_i - (n_1/n_2)^2}}. \tag{8.18}$$

For a fixed beam geometry, the penetration depth is seen to vary in inverse proportion with the frequency, Figure 8.23. The penetration depth ranges between a few and a few hundred micrometres.

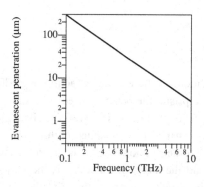

Figure 8.23 Penetration depth of an evanescent wave. The calculation is for refractive indices $n_1 = 1$, $n_2 = 3.5$ and angle of incidence $\theta_i = \pi/4$.

Two applications of the evanescent wave emanating from the hypotenuse of a prism are shown in Figure 8.24.

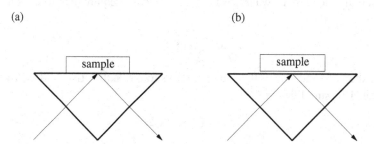

Figure 8.24 (a) Attenuated total reflection. The sample is touching the prism surface (Kretschmann geometry). (b) Frustrated total internal reflection. The sample is very close to the prism surface (Otto geometry).

Attenuated total reflection

If a sample is placed on the surface of the prism, Figure 8.24a, the evanescent wave can penetrate the sample slightly, the result being that the wave is not 'totally' reflected but suffers some attenuation. This method is called **attenuated total reflectance**. It is useful for measuring the absorption in highly absorbing materials, with absorption coefficients in the range 1 to $100\,\text{m}^{-1}$. Since the penetration is greater at lower frequencies (Figure 8.23), this method is more sensitive to absorption at lower frequencies. Compared to normal transmission methods, it is difficult to make the attenuated total reflectance method quantitative.

Frustrated total internal reflection

If a sample is placed slightly beyond the surface of the prism, Figure 8.24b, the evanescent wave can penetrate the gap and then continue into the sample. Again, the result is that the wave is not 'totally' reflected. In this case, the reflection has been *frustrated* by the layer of less dense material between the prism and the sample. This method is called **frustrated total internal reflectance**. It is used in studying surface plasma resonances.

8.6 Attenuators

Attenuators are used to reduce the radiation in the beam path. In the terahertz regime, the problem is usually too little radiation, rather than too much. So attenuators are not used as extensively as, say, in the visible range.

Often the intensity of the terahertz beam may be controlled in the source itself, for example, by changing the bias on a photoconductive emitter.

We have spoken often of the absorption characteristics of many materials, and how this renders them unsuitable for lenses or prisms. This property does, however, make them suitable to attenuate terahertz beams. Materials such as PE or PTFE can be placed in the beam, a known thickness producing a known attenuation. Even sheets of paper can be used. The attenuation can be quite precise if the terahertz beam is monochromatic.

Broadband and calibrated attenuators may be manufactured using a thin amount of metallisation on a substrate such as silicon. An alternative is to use a metal mesh.

8.7 Filters

A filter cuts out some frequencies of light and allows others to pass. Three ideal **filters** are shown in Figure 8.25.

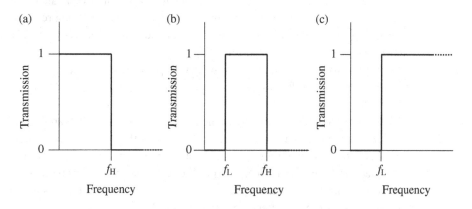

Figure 8.25 Ideal filters. (a) Low-pass filter. (b) Band-pass filter. (c) High-pass filter.

An ideal **low-pass filter** transmits all light below a certain frequency (f_H) and blocks all light above that frequency. An ideal **high-pass filter** transmits all light above a certain frequency (f_L) and blocks all light below that frequency. An ideal **band-pass filter** transmits all light in a certain frequency range, the pass band ($f_L < f < f_H$), and blocks all light outside that frequency range, the stop bands ($f < f_L, f > f_H$). A *band-stop filter*, also called a *notch filter*, not shown in Figure 8.25, is the inverse of the band-pass filter, stopping light in a certain frequency range but transmitting all light outside that range.

Since they transmit radiation with long wavelengths, the low-pass filter is also called the *long-pass filter*. Likewise, the high-pass filter is called the *short-pass filter*. Be aware of the difference; it is not always clear whether it is frequency or wavelength that is being used to specify the filter.

The most commonly used filters in the terahertz regime are low-pass filters. These are often used to transmit the terahertz radiation but remove unwanted higher-frequency radiation. For example, near-infrared and visible radiation is prevented from reaching a detector to reduce the load on the detector and increase its sensitivity. Another application is to remove higher harmonics, at $2f, 3f, 4f \ldots$ from a diffraction spectrometer designed to operate at frequency f. A commonly used low-pass filter is black polyethylene. This material is readily available as agricultural plastic or garbage bags.

High-pass filters, in the context of terahertz applications, block lower-frequency radiation, such as in the microwave, TV and radio region. These are not widely employed.

Band-pass filters may be based on interference effects, as in the visible. They may be fabricated, for example, using stacks of semiconductor plates. Band-pass filters may also be used by exploiting the reststrahlen radiation from ionic crystals. A third way to make a band-pass filter is to use a metal mesh. Two types of metal mesh are used: the normal mesh, the usual mesh, where thinner wires separate large gaps, and the *anti-mesh*, where thinner spaces separate large areas of metal. (The anti-mesh can be made by depositing metal blocks on a transmitting substrate.) The mesh is modelled by an inductive equivalent circuit and the anti-mesh by a capacitative equivalent circuit. When the wavelength of radiation is the same as the mesh period, a strong resonance occurs. The mesh, or inductive grid, transmits almost all radiation at the resonant frequency. The anti-mesh, or capacitative grid, reflects almost all radiation at the resonant frequency. So, in transmission, the mesh acts as a band-pass filter and the anti-mesh acts as a band-stop filter. (In reflection, the mesh is a band-stop filter and the anti-mesh a band-pass filter.)

Some filters operate by absorbing the unwanted radiation. They are therefore known as *absorption* filters. They are convenient to use, as they may be placed directly in the beam, and the unwanted radiation is collected. However, in the process of capturing the unwanted radiation, the filter heats up. The filter may then re-radiate as a blackbody source at a higher temperature, which is generally undesirable. For this reason, in some applications, absorption filters are actively cooled.

Another type of filter is one based on *scattering*. An example is polystyrene, a common packing material. This will transmit terahertz radiation, but scatter visible radiation. Such filters may be inexpensive and easy to insert in the optical beam, but bring the risk that the final destination of the unwanted, scattered radiation may be difficult to determine.

Another type of filter again is based on *reflection*. These may be tuned, for example, by polishing a metal to different degrees of roughness, or by making wire grids with different spacings between the wires, as mentioned above, or reststrahlen filters, also mentioned above. One difficulty in using reflective filters is that the optical path is changed. Two reflective filters are needed to return the beam to be parallel to its original direction and three needed to return the beam to its original path.

8.8 Windows

In everyday life, a window is an opening that lets light in. Often it is covered with glass, which has the dual purpose of letting light in, but keeping other things – noise, dust, insects – out.

In terahertz applications a window is not a simple opening but an opening covered by a material that will transmit terahertz radiation but keep other things out. Most commonly, the window is part of a vacuum system, and separates air on the outside from the evacuated space on the inside.

A terahertz window may then be regarded as a subset of the terahertz filters. As a filter, it must pass the terahertz radiation. But it must have in addition mechanical properties to block out whatever is required; for example, it must be strong enough to stand a pressure difference of one atmosphere across its faces if used as a vacuum window. Some materials that serve well as optical filters, such as wire meshes and polystyrene, will not do as vacuum windows, as they are porous.

Windows may be made out of crystals or polymers. Many I-VII compounds are suitable, such as LiF, NaCl, KCl, KBr, KI, CsBr, RbBr; many of these absorb atmospheric water, and this must be guarded against in use. Likewise, II-VI compounds, such as ZnS, ZnSe, CdS, CdTe; III-V compounds, such as GaAs and InAs; and group IV elements, C in the form of diamond, Si and Ge may be used. It is important that the semiconductor compounds have a low level of doping, or they become metallic and so opaque; silicon can be prepared to a very high quality and high-resistivity silicon is often used in terahertz optics. Other crystal window materials are sapphire and crystalline quartz (cut along the Z axis to avoid birefringence). Suitable polymers include polyethylene (PE), both high-density (HDPE) and low-density (LDPE), polytetrafluoroethylene (PTFE or teflon), polypropylene (PP) and polymethylpentene (PMP or TPX). TPX is notable for being rigid and easily machined and transparent in the visible. Glass, as used for windows in the visible, absorbs strongly at terahertz frequencies, so is not suitable as a terahertz window.

8.9 Polarisers

Apart from direction and optical fluence, which mirrors and lenses can change, and spectral composition, which filters can change, it is desirable to control the polarisation of terahertz radiation. Devices that permit this are called **polarisers**.

8.9.1 Linear polarisers

Linearly polarised radiation is most simply produced by wire grids.

The wire grid is typically made by winding a high-tensile metal, like tungsten, onto a frame. The wire used is very thin, typically 50 μm in diameter. The spacing between the wires is very small, typically 100 μm. This is a **free-standing grid**.

Alternatively, a metallic grid may be deposited on a transparent substrate. Typically, polyethylene is used as a substrate and aluminium is the deposited metal.

Light polarised along the direction of the wires, $E_\parallel$, sets the electrons in the wire in motion; they reradiate the light back to where it came. You can think of the grid as acting as a perfect metallic reflector.

Light polarised perpendicularly to the direction of the wires, $E_\perp$, does not set the electrons moving. The radiation is not absorbed or reflected but passes straight through. The grid acts as a perfect transmitter.

For an arbitrary angle, θ, between the grid direction and the electric field, the transmitted intensity, I, varies as

$$I = I_0 \sin^2 \theta, \qquad (8.19)$$

where I_0 is the intensity falling on the polariser. $E_\parallel$ corresponds to $\theta = 0$ and $E_\perp$ corresponds to $\theta = \pi/2$. (You may have seen a simple picture of mechanical polarisation where a rope is shaken through a picket fence. This gives completely the wrong picture for the relationship between the plane of polarisation of light and the direction of the metallic grid.)

The extinction ratio of a polariser is defined as the ratio of wanted polarisation to unwanted polarisation intensity. The extinction ratio is $I_\perp/I_\parallel$. Ideally, it should approach infinity; in practice, it is in the range of around 100 to 1000.

The grid spacing determines the highest frequency at which the polariser may be used. Once the wavelength is smaller than about the grid spacing, the light can leak through the polariser, so the polariser can be used as a high-pass filter. As the grid spacing decreases, the extinction ratio increases (but at the expense of decreasing the lowest frequency at which the polariser can be used).

The dual property of the wire grid – transmitting one polarisation, reflecting the other – means it can be directly employed as a (polarising) beamsplitter.

A completely different way to make a terahertz polariser is using the polarising properties of reflection at the Brewster angle. Rather than one reflecting surface, the effect is increased by using several in series, and the resulting unit is called a *pile of plates*. Such polarisers do not have the high extinction ratios of wire-grid polarisers and so are not widely used.

8.9.2 Wave plates

To produce elliptically polarised light, and circularly polarised light (a special case of elliptically polarised light), **wave plates** are used. Wave plates may be prepared in the terahertz regime as in the visible regime by using a birefringent crystal cut along the appropriate axis to the appropriate thickness. Given that the wavelength of terahertz radiation is so much greater than that of visible light, if the birefringence is similar, terahertz wave plates will be thicker and simpler to manufacture. A wave plate prepared in this way will suffer the same restriction in the terahertz as in the visible, namely, that it only operates correctly at one frequency.

A typical material used for terahertz wave plates is crystalline quartz (x-cut). Around 1 THz, the difference in the refractive indices of the extraordinary and ordinary rays is about $n_E - n_O = 0.04$.

The usual development of the theory to obtain the requisite thickness follows. Traversing a distance d in a direction with refractive index n_O for the ordinary ray corresponds to an optical path length of $n_O d$. Traversing the same distance corresponds to an optical path length $n_E d$ for the extraordinary ray. The optical path length difference is

$$\Delta = (n_E - n_O)d. \tag{8.20}$$

The corresponding phase difference is

$$\delta\phi = \frac{2\pi f}{c}(n_E - n_O)d. \tag{8.21}$$

If the phase delay is π, we have a **half-wave plate** or $\lambda/2$ plate. Such a plate has the property of rotating the axis of polarisation of linearly polarised light falling on it through twice the angle that the plate makes with the direction of polarisation. Regarding circularly polarised light, transmission through the half-wave plate reverses the sense of rotation.

If the phase delay is $\pi/2$, we have a **quarter-wave plate** or $\lambda/4$ plate. Such a plate has the property of converting circularly polarised light to linearly polarised light and vice-versa. Either sense of rotation of the circularly polarised light may be produced.

Liquid crystals are useful in this respect as their birefringence may be tuned. The phase delay can be controlled electrically.

Another scheme is to couple a wire-grid polariser with a mirror. Changing the grid-to-mirror distance allows the phase delay to be controlled.

8.10 Summary

8.10.1 Key terms

geometrical optics, 162	lenses, 174	windows, 185
quasi-optics, 163	prisms, 180	polarisers, 186
mirrors, 166	filters, 183	wave plates, 187

8.10.2 Key equations

Gaussian beam	$w(z) = w_0 \left[1 + \left(\dfrac{cz}{\pi w_0^2 f} \right)^2 \right]^{1/2}$	(8.3)
Rayleigh range	$z_R = \dfrac{1}{2} k w_0^2 = \dfrac{\pi w_0^2}{c} f$	(8.5)
critical angle	$\theta_c = \arcsin \dfrac{n_1}{n_2}$	(8.15)
polariser transmission	$I = I_0 \sin^2 \theta$	(8.19)
wave plate phase delay	$\delta\phi = \dfrac{2\pi f}{c} (n_E - n_O) d$	(8.21)

8.11 Table of symbols, Chapter 8

General mathematical symbols appear in Appendix B. If the unit of a quantity depends on the context, this is denoted '—'.

Symbol	Meaning	Unit
a	grating period	m
a	semimajor axis of ellipse	m
a	parabola parameter	—
a	Gaussian parameter	m^{-2}
b	semiminor axis of ellipse	m
c	lightspeed	m Hz
d	distance	m
d_a, d_b	lateral distance; incoming, outgoing	m
E	electric field amplitude	V m^{-1}
E_0	electric field amplitude, initial	V m^{-1}
f	frequency	Hz
f_H, f_L	frequency; high, low	Hz
H	offset from sphere centre	m
I	electric field intensity	J m^{-2} Hz
I_0	electric field intensity, initial	J m^{-2} Hz
J	distance of focus from sphere centre	m
L_p, L_r	parent, reflected focal length	m
n	refractive index	[unitless]
n_1, n_2	refractive index; first, second medium	[unitless]
n_E, n_O	refractive index; extraordinary, ordinary ray	[unitless]
r	distance from z axis	m
R	radius of sphere	m
s_i	distance of image	m
T	height of truncated sphere	m
w, w_0	width of waist, minimum width of waist	m
X	hyperbolic parameter	m
x, y, z	coordinate labels	—
z_R	range, Rayleigh	m
Z	hyperbolic parameter	m
α, β	entrance, exit angle	rad
δ	phase difference	rad
Δ	path difference	m
λ	wavelength	m
θ_0	half divergence angle	rad
θ_1	angle from normal, first medium	rad
θ_2	angle from normal, second medium	rad
θ_c	incidence angle	rad

9 DETECTORS

The mathematics used in this chapter is limited to the operations of addition, subtraction, multiplication, division and a little differentiation.

In this chapter, we look at sensors of terahertz-frequency electromagnetic radiation. Terahertz sensors detect invisible terahertz radiation and convert it to something perceptible to a human being – usually a number on a dial or on a screen.

Terahertz-frequency electromagnetic radiation is invisible. The human eye cannot see it. How can we tell it is there?

The other human senses are not much help. We cannot taste, or hear, or smell terahertz radiation. We can feel it – we perceive terahertz radiation as heat – but only if it is rather intense. The human body, on the whole, is rather insensitive to terahertz radiation.

Devices that are sensitive to terahertz radiation and convert it to a signal that humans understand are called *sensors* or *detectors*. The general term for a device that converts a signal from one form to another is a *transducer*. Most commonly, the transducer takes a physical property, such as temperature, acceleration or viscosity, and converts it to an electrical signal, since electrical signals are easily stored and manipulated. Of course, a terahertz-frequency electromagnetic wave is already an electrical signal, but it is at too high a frequency and often at too low an intensity for a human to sense easily. A terahertz sensor takes in the terahertz-frequency electromagnetic wave and gives out or records a reading at a lower frequency that is perceptible to you or me.

That terahertz-frequency radiation may be perceived as heat segues into one type of terahertz sensor, the *thermal sensor*. Thermal sensors are also called *calorimetric sensors*. In a thermal sensor, the terahertz radiation heats a detector element, causing the temperature of the element to rise. The increase in temperature is measured by a change in a thermometric property of the element, such as its volume, its electrical resistance or the electrical charge on its surface. Changing a thermometric property is the principle of operation of the *Golay cell*, the *bolometer* and the *pyroelectric detector*.

Terahertz radiation falling on an electro-optic material changes the birefringence of the material. The change in birefringence in turn may be monitored by measuring the rotation of the polarisation of visible or near-infrared light traversing the material. This is the principle of operation of the *electro-optic detector*.

Terahertz radiation striking a material containing mobile charge carriers can affect the current flow. This is the basis of the *photoconductive detector*.

The three main types of terahertz sensor are related to three ways of producing terahertz radiation. There are thermal, electro-optic and photoconductive sensors, just as there are thermal, electro-optic and photoconductive emitters. Although the processes involved in sensors and sources are not precisely the reverse of each other, some insight into the operation of sensors comes from a knowledge of sources, and vice versa.

Learning goals

After studying this chapter, my expectation is you will appreciate

- the key parameters that characterise detectors of terahertz radiation,
- the physical principle of operation of thermal detectors,
- the operation of detectors of pulsed terahertz radiation.

9.1 Detector parameters

Asking 'Which is the best detector?' is akin to asking 'Who is the best athlete?' A top sprinter is unlikely to be an excellent marathon runner: the physical requirements for the two events are quite different, explosive power compared to endurance. The profile of a decathlon champion is different again. Likewise, different sensors have different strengths and weaknesses. The question of which is the best sensor can only be answered in the context of the application.

So, what properties or characteristics of a sensor are important?

Sensitivity The sensitivity of a transducer is the ratio of the output to the input. For a sensor that gives an output as an electrical signal, V, measured in volts, for a given radiation power, P, measured in watts, the sensitivity will have units of V/W. In this case, the sensitivity is called the voltage responsivity, defined by

$$R_V \equiv \frac{dV}{dP}. \tag{9.1}$$

Alternatively, if the output is in the form of a current, the sensitivity is expressed in amps per watt. Generally speaking, high sensitivity is desirable. This is an important characteristic, but perhaps not the most important, as the limit to what can be measured is expressed in the detectivity, not the sensitivity.

Linearity Ideally, we want a detector to have the same sensitivity regardless of the size of the input signal. For example, if a terahertz signal of power 1 W produces an output of 2 V, we would like a terahertz signal of five times more power, 5 W, to produce an output of five times more, 10 V. The extent to which the sensor sensitivity varies with input power is termed the nonlinearity.

Detectivity Arguably the most important characteristic of a sensor. Detectivity refers to the smallest signal that can be sensed. In many cases, this limit is set not so much by the sensor being unable to respond to the terahertz-frequency radiation as by other spurious signals confounding the signal of interest. Unwanted signal is called noise. The limit to the detectivity is sometimes expressed as the size of the wanted signal to the unwanted signal, termed the signal-to-noise ratio. It is better expressed in terms of noise equivalent power, the power corresponding to the noise of the sensor.

Range The range is the bookend to detectivity. While detectivity refers to the minimum signal the sensor may handle, range specifies the maximum signal that the sensor can suffer before damage results.

Speed When a signal is suddenly turned on, some detectors will be quicker than others to respond to the change, and register the new value of the terahertz field. This property may be quantified by such parameters as the response speed or the time constant. The property is also referred to as the *frequency response*, but I will avoid this term, as it may be confused with the sensor's response to various frequencies of radiation, which I call the spectral response.

Spectral response So far I have been referring simply to the terahertz signal, but of course a terahertz wave has properties other than simply its intensity. A key characteristic is its frequency. For example, a sensor may be excellent at detecting waves of 1 THz frequency, but very poor at detecting waves of 2 THz frequency. Generally, we prefer a sensor that detects signals over a broad band of frequencies (a broadband detector), but in some applications we may prefer to detect only some frequencies and reject others (a narrowband detector). Generally speaking, we prefer the sensitivity to be the same at all frequencies. A specification of the detector sensitivity at different frequencies constitutes the spectral response.

Polarisation Many sensors are equally sensitive to radiation of any polarisation – circular, elliptical or linear. For a particular purpose it may be desirable to have a detector that is responsive, for example, to one particular plane of linear polarisation. It is usually undesirable to have a detector whose polarisation response varies with frequency, but this is a property of some detectors.

Environmental factors Some sensors reside in controlled conditions in a laboratory; others find themselves in more difficult environments, such as on a satellite in earth orbit. Depending on the application, the sensor's size, weight, robustness and temperature and pressure requirements may be important. The ideal is to have a compact, light, and robust sensor that operates at ambient temperature and pressure.

Let's look into the concept of detectivity a little more closely. Associated with this is the term noise equivalent power (NEP).

The idea behind NEP is a simple one. With no radiation at all falling on the sensor, there is still some signal at the sensor output. This is sometimes called the background signal. This unwanted, but unavoidable, background signal is often referred to as noise.

Now, let's say a small amount of radiation falls on the sensor, amounting to, say, one-tenth of the background noise. The small change in the output of the sensor with the terahertz radiation relative to without the terahertz radiation will be difficult to detect (although it is possible). Let's now consider a large amount of radiation falling on the sensor, say, ten times the background noise. This will be very easy to detect. Let's now consider the terahertz power that needs to fall on the sensor so that the signal at the sensor output is exactly the same size as the signal produced by the background noise. This amount of power is the NEP. It is measured in watts.

Let's say we have a detector of responsivity R_V. (Remember, R_V is simply the ratio between the change in output voltage and the change in incident power.) If we denote by V_{noise} the voltage corresponding to the background noise, we have

$$\text{NEP} = \frac{V_{\text{noise}}}{R_V}. \tag{9.2}$$

Generally speaking, it is desirable to have a low NEP. This means that the noise in the detector corresponds to a low power of terahertz radiation.

To precisely define the NEP, we need to precisely define the conditions of the measurement. So the NEP is usually defined for a specific frequency, say, 1 THz. In practice, it is impossible to measure at exactly and only that frequency; we measure over a range of frequencies, say 0.9 to 1.1 THz, or over a frequency bandwidth, in this case 0.2 THz. There is a further consideration. Different detectors operate at different speeds, so the speed at which the measurement is made needs to be taken into account if a fair comparison is to be made. For this, we assume that the terahertz signal is being turned on and off (modulated) at a given frequency, say at 1 Hz. (Don't confuse the modulation frequency with the frequency of the terahertz radiation itself.)

You might be familiar with the signal-to-noise ratio (SNR) and wonder why NEP is used in preference. If the noise is independent of the power, the SNR depends on the experimental arrangement, whereas the NEP does not, but rather is characteristic of the sensor.

Example 9.1   SNR and NEP

This example illustrates how SNR may vary with experimental conditions while NEP remains the same.

A new terahertz sensor is being tested against a calibrated terahertz source. When the source is turned off, the sensor produces an output of 1 V.

(a) When the source emits radiation of power 1 nW, the sensor produces an output of 3 V. What is the SNR? What is the NEP?

(b) When the source emits radiation of power 2 nW, the sensor produces an output of 5 V. What is the SNR? What is the NEP?

Solution

To begin with, when there is no radiation falling on the sensor, the signal it produces is 1 V. So the noise is 1 V.

(a) The additional signal produced under 1 nW of terahertz radiation is 3 V − 1 V = 2 V. The ratio of signal to noise is then (2 V/1 V) = 2. (We have assumed that the noise has not changed.) To calculate the NEP, we need to know what power corresponds to a signal of 1 V. Since a change in power of 1 nW corresponds to a change in output of 2 V, we deduce a change in output of 1 V corresponds to 0.5 nW. Hence the NEP is 0.5 nW.

(b) Proceeding as in part (a), the signal is now 4 V, the noise remains 1 V and so the SNR is 4. Since a change in output of 4 V corresponds to an input of 2 nW, we deduce the input power corresponding to 1 V at the output is 0.5 nW, as before.

So, as the experimental conditions are varied the SNR changes, but the NEP does not.

Exercise 9.1 Output

The responsivity of a Golay detector is stated to be 10^5 V/W. What output is expected when 10 µW of power is incident on the detector? ∎

Exercise 9.2 Responsivity

An input signal of strength 10 µW generates a potential difference of 1 µV at the output of a thermopile detector. What is the responsivity of the thermopile, expressed in mV/mW? ∎

Exercise 9.3 Linearity

A thermopile is found to have a responsivity of 0.1 mV/mW when sensing powers below 1 mW. A stronger source, of 100 mW, produces a response of 10 V at the detector output. Is the detector linear? ∎

So far, so good. Now we come to a possible point of confusion. Sometimes, the NEP is defined differently from how I have just defined it. In the given definition, we assumed certain conditions would be fixed in specifying the NEP: the frequency of the radiation to be detected, the modulation frequency and the bandwidth. Some prefer to have an 'NEP' that is independent of the bandwidth. The reasoning is as follows. The longer the time that can be spent taking a measurement, the lower the noise will be, and also the smaller the bandwidth will be. So there is some virtue in comparing sensors using a standard measurement time or, equivalently, a standard bandwidth. A further complication is now introduced. Under the assumption that the square of the noise power varies with the bandwidth, rather than compare noise per bandwidth, it is more reasonable to compare noise per root bandwidth. So we can define a noise equivalent power per root bandwidth $NEP_{prb} = NEP/\sqrt{\Delta f}$. The confusion arises that the term NEP is used by some to refer to NEP_{prb}. Note that NEP has units of power (watts), which makes sense; in contrast, NEP_{prb} has units of watts per root hertz.

Other ways of dealing with the same general concept are in use. Some define the detectivity to be the reciprocal of the NEP:

$$D \equiv \frac{R_V}{V_{noise}} = \frac{1}{NEP}. \tag{9.3}$$

So far, we have ignored the physical size of the sensor. It might be expected that a larger sensor element would produce a larger signal than a smaller one, and this might be incorporated into the comparison. This leads to the concept of specific detectivity, or detectivity scaled for detector size and bandwidth, written as D^* and pronounced as 'D-star'.

Exercise 9.4 Johnson noise

Johnson noise, or thermal noise, appears across a resistance R at temperature T as a voltage $V = \sqrt{4k_{\mathrm{B}}T\Delta f R}$ for a bandwidth Δf. Estimate the Johnson noise for a detector of 2 MΩ resistance operating near room temperature (295 K) with a bandwidth of 10 Hz. If the responsivity of the detector is 100 kV/W, what is the NEP under these conditions? How does the NEP change if the detector is cooled to 4 K but the other factors remain the same? ∎

Exercise 9.5 D-star

Calculate D^* for the detector in the previous exercise at both 4 K and 295 K. ∎

9.2   Thermal detectors

A **thermal detector** is heated by the radiation falling on it. The received heat results in an increase in temperature. The temperature rise is reflected in a physical change in the detector, such as the detector expanding. The main features of the thermal detector are shown in Figure 9.1. In thermal equilibrium, the radiant power, P, being absorbed,

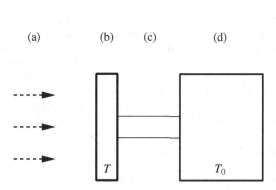

Figure 9.1 A thermal detector. (a) Radiation enters from the left. (b) The absorber. A perfect absorber will trap all radiation striking it. The energy from the radiation heats the absorber and raises its temperature to T. The absorber has thermal capacity C. (c) A thermal link connects the absorber to the thermal reservoir. The heat link has thermal conductance G. (d) The thermal reservoir, or heat sink, is defined purely by its temperature, which is always T_0.

raises the temperature of the absorber, T, above the temperature of the thermal reservoir, T_0, by an amount that depends on how rapidly the heat is conducted away by the heat link, of conductance G. We have

$$P = (T - T_0)G. \tag{9.4}$$

Provided we know the values of all the quantities on the right-hand side of the equation, we can determine the radiant power on the left.

If, rather than equilibrium, we have a changing power level on to the absorber, the absorber temperature will fluctuate. The pertinent parameter is the thermal time constant $\tau_{thermal}$,

$$\tau_{thermal} = \frac{C}{G}, \qquad (9.5)$$

where C is the thermal capacitance of the absorber.

These general principles apply to all thermal detectors. Let's now look at three specific examples.

9.2.1 Thermocouple and thermopile sensors

The **thermocouple** and **thermopile** are very simple thermal detectors. The essential components are shown in Figure 9.2. The thermocouple, as might be inferred from the

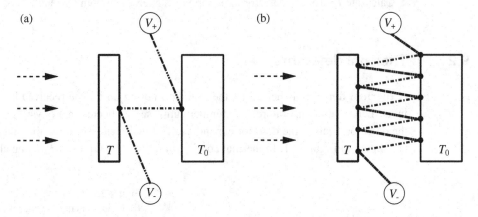

Figure 9.2 (a) A thermocouple is made from two – a couple of – dissimilar materials. Usually the two materials are two metals. Here one metal is represented by a dash/dot line and the other by a double-dash/triple-dot line. Wires made from the two metals are joined at a point and this joint is thermally connected to a reservoir at temperature T_0. This is the *cold junction*. One of the wires coming from this joint is connected to the terminal of a voltmeter, here shown as V_+. The other wire is used to form a second junction with a second piece of the second wire; this joint is at temperature T and called the *hot junction*. The loose end of wire is connected to the second terminal of the voltmeter, here shown as V_-. A temperature difference between the hot and cold junctions will result in a potential difference being indicated by the voltmeter. The larger the temperature difference, the greater the magnitude of the voltmeter readout. (b) A thermopile is made up of many thermocouples joined one to another. The voltage produced across the ends is multiplied by a factor equal to the number of thermocouples in the thermopile.

name, is made up of a couple, or two, pieces of metal. The thermopile, again as might be inferred from the name, is made up of a pile, or stack, of thermocouples.

In the thermocouple, one joint between dissimilar metals is located at the thermal reservoir (the *cold junction*) and another is located at the absorber (the *hot junction*). The potential developed between the two joints increases as the temperature difference does.

In the thermopile, the potential developed is multiplied by the number of thermocouples involved. The thermopile is thus more sensitive than the thermocouple, but at the expense of a greater thermal mass and thermal conductance.

9.2.2 Pneumatic – Golay sensors

The **Golay cell** was invented by – you guessed it – Golay.

The physical principle of operation of the Golay cell is shown in Figure 9.3. Terahertz-frequency radiation falling on a small bag of gas heats the gas, causing it to expand. The change in volume of the gas cell is monitored, usually by a beam of light bouncing off a reflecting membrane, and the radiation intensity deduced accordingly. Of course, there is more to it than that, since the gas bag must be allowed to cool before it can be used again, and there is a delicate interplay between the thermodynamic parameters that is controlled to optimise this.

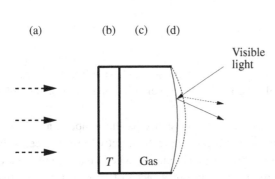

Figure 9.3 The Golay or pneumatic cell. (a) Radiation enters from the left, typically through a window (not shown). (b) The radiation heats up the absorber element, usually a thin film. (c) The heat is transferred to a volume of gas. A monatomic gas, such as argon, is normally used. (d) As the gas is heated, it expands. This causes the deflection of a thin membrane at the end of the gas cell. Reflection of visible light from the flexible membrane is used to measure the deflection.

The sensitivity of the Golay cell is usually of the order of 10^5 V/W. This is a range that makes it convenient to measure the radiation from typical laboratory thermal sources.

A convenience of the Golay cell is that it operates at room temperature. (In contrast, for example, the bolometer requires cooling, typically using liquid helium.) A limitation of room-temperature operation, however, is that the thermal excitation of the sensor itself at room temperature then sets the limit of detectivity.

Golay cells are relatively compact, typically of dimensions $2 \times 10 \times 15$ cm^3, simple to use and find widespread applications in laboratories.

Their main limitations are fragility – sensitivity to vibration and to damage by too strong radiation; relatively slow response, of the order of one second; and the limitation in detectivity set by room-temperature operation.

9.2.3 Pyroelectric sensors

Certain crystals sustain an electric field across opposite faces, the strength of which depends on the temperature of the crystal. Such crystals are termed *pyroelectric*. As terahertz-frequency radiation falls on a pyroelectric crystal and heats it, the electric field across the face will change, permitting the strength of the radiation to be gauged. This is the principle of operation of the **pyroelectric detector**, Figure 9.4.

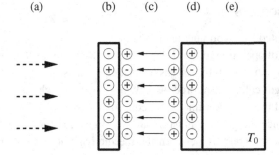

Figure 9.4 Pyroelectric detector. (a) Radiation enters from the left, through a transparent front electrode (b). The pyroelectric material (c) acquires a spontaneous polarisation and surface charge accumulates. The back electrode (d) separates the pyroelectric element and the heat sink (e).

Pyroelectric sensors are usually compact and robust. These are their strengths. However, the sensitivity is low, the response speed only moderate (tenths of a second) and the detectivity limited by the room-temperature operation. These are their weaknesses.

9.2.4 Bolometric sensors

The word **bolometer** means, literally, heat measurer. Most bolometers measure heat by a change in their electrical resistance. The principle of operation of the bolometer is shown in Figure 9.5.

Every material changes its electrical resistance as the temperature changes, some more so than others. In many cases the change in electrical resistance, at least to a first approximation, is directly proportional to the change in temperature. For example, doubling the thermodynamic temperature of a piece of copper wire, say from room temperature (~300 K) to 600 K, results in the resistance approximately doubling.

In semiconductors, however, the change of resistance on heating is rather more spectacular. Or should we better say the change on cooling – for as a semiconductor is cooled, its resistance increases (in contrast to the copper, whose resistance decreases) and at an exponential rate (in contrast to the linear rate for the copper wire). The reason for this is that as the semiconductor is cooled, the charge carriers *freeze out*. As the temperature is lowered, intrinsic charge carriers return to the band from which they came, conduction or valence. Indeed, at zero temperature, a semiconductor is an insulator. At low temperatures – a few degrees above absolute zero – only very few electrons can be promoted across the bandgap by the available thermal energy, so the resistance is very high. As the temperature increases, the number of electrons promoted across the bandgap increases exponentially, and the resistance drops accordingly. The differential change of resistance with temperature is greater at lower temperatures.

A semiconductor bolometer therefore increases in sensitivity as the temperature decreases. Such bolometers are typically operated at the boiling point of liquid helium,

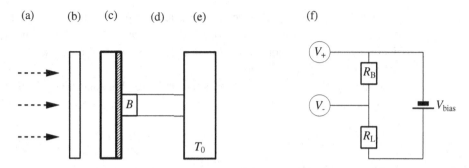

Figure 9.5 Bolometer. (a) Radiation enters from the left. Typically, it is concentrated on to the bolometer element by a cone (not shown). (b) A cold window is often used to remove unwanted high-frequency radiation from the bolometer element to increase its sensitivity. (c) The bolometer element. A composite bolometer is shown. A thin metal film, such as bismuth, shown here by shading, acts as the absorber. It is supported by a transparent substrate, such as diamond. The sensing element (B) changes resistance rapidly with temperature. Silicon, germanium, or carbon are typically used. (d) The heat link connecting the sensing element to the heat sink (e). (f) Electrical circuit of the bolometer. A battery supplies a steady bias across the sensing element in series with a load resistor. As radiation heats the sensing element, its resistance R_B decreases. This reduces the voltage drop across the element. Usually the load resistor is cooled, to decrease the thermal noise in it. Typically the sensor element is at a temperature of 4 K, the boiling temperature of liquid helium at room temperature. By reducing the pressure over the boiling helium, operation may be achieved at even lower temperatures.

4.2 K at atmospheric pressure. By reducing the ambient pressure (by removing the evolved helium gas with a vacuum pump) even lower temperatures, typically 2 K, and so even higher sensitivities, are reached.

The necessity of operating at low temperature brings with it an advantage – the thermal vibrations in the detector itself (in the absence of terahertz radiation) are much reduced compared to a room-temperature detector. The detectivity is consequently much superior.

Many bolometers are cooled using liquid helium. There is an overhead connected to this: filling takes time, the helium boils away and needs to be replenished, the consumption of helium costs money. An alternative is to cool with a closed-cycle refrigerator. This is more convenient and cost-effective in that liquid helium does not have to be supplied and replenished, but the requirement to have a helium compressor adds to the bulk and the noise of the installation.

Since the changing temperature is detected not by a bulk change in the detector element, such as thermal expansion, but by changing the electron population, the bolometer is very fast.

The usual materials used for bolometer sensors are the elemental semiconductors silicon and germanium, but other materials can be employed, for example, a carbon resistor.

As with the other thermal detectors, the detector element has to be cooled again before it can detect subsequent radiation. There is a tradeoff between sensitivity, which is increased by a weak connection to the cold reservoir, and response time, which is improved by strong thermal contact.

In summary, bolometers are fast, sensitive and boast high detectivity. With the necessity to cool to within a few degrees of absolute zero, they are relatively cumbersome and bulky.

9.3 Electro-optic detectors

The electro-optic effect has been discussed previously in relation to the generation of terahertz radiation. Here the inverse effect is employed to permit the detection of terahertz radiation.

A property of an electro-optic material is that the presence of a terahertz field changes the refractive index at visible frequencies. This change in refractive index may be monitored by a visible light beam and related to the terahertz radiation falling on the material.

More specifically, the change of refractive index for light of one polarisation (the ordinary, or o wave) is different from the change of refractive index for light of the orthogonal polarisation (the extraordinary, or e wave). It is this polarisation difference that is most conveniently measured.

So an **electro-optic detector** comprises many separate components, as shown in Figure 9.6. Apart from the electro-optic crystal, a probe beam must be supplied, and its polarisation rotation measured, typically via a quarter-wave plate, a Wollaston prism and a pair of photodetector diodes feeding into a differential amplifier.

In choosing an electro-optic crystal, several factors come into play. First, a large electro-optic coefficient is desirable. Next, the crystal must be matched for phase coherence for the two frequencies required: the terahertz frequency being detected and the probe frequency being employed (so, for example, for detection using 800 nm probe radiation, ZnTe is often employed; at 1560 nm, GaAs is preferred). Finally, a crystal with a high damage threshold is preferred over one with a low damage threshold.

There are several practical matters to be taken into account in setting up an electro-optic crystal. First, the crystal should be cut so that the electro-optic response is as high as possible. The (110) crystallographic direction is usually chosen for this purpose. Second, the thickness of the crystal needs to be optimised for the application. A thicker crystal will give a greater rotation of the polarisation and so a stronger output. However, a thinner crystal will give better phase matching and so greater bandwidth. There is therefore a tradeoff between the sensitivity and the bandwidth. Third, the crystal needs to be oriented in the best way for the polarisation of the terahertz radiation to be detected. For unpolarised radiation, this is irrelevant. For linearly polarised radiation, it is crucial. For example, for a (110) oriented crystal, there are four orientations at which the signal from linearly polarised radiation is at a maximum.

A main advantage of the electro-optic detector over the thermal detectors is that the detection of radiation is coherent. The terahertz signal is detected precisely when the

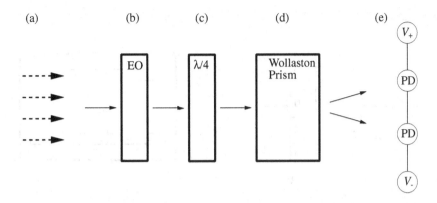

Figure 9.6 Electro-optic detection. (a) The terahertz pulse enters from the left (dashed arrows). A synchronised, pulsed probe beam of visible or near-infrared radiation is directed along the same path (full arrows). The probe radiation is linearly polarised. (b) The terahertz radiation and the probe radiation together enter an electro-optic crystal, such as ZnTe. In the electro-optic crystal, the effect of the terahertz field is to rotate the polarisation of the probe beam. (c) A quarter-wave plate (for the wavelength of the probe beam) is used to assess the polarisation of the probe beam. In the absence of terahertz radiation, the quarter-wave plate produces circularly polarised light. The effect of terahertz radiation is to change this to elliptically polarised light. (d) The Wollaston prism is to continue the assessment of the polarisation rotation of the probe beam through the electro-optic crystal. The Wollaston prism separates the elliptically polarised light into two, perpendicular, linearly polarised beams. In the absence of terahertz radiation, the probe beam before the prism will be circularly polarised and after the prism will be split into two beams of perpendicular polarisation and equal intensity. (e) The two beams from the prism are detected by two photodiodes in a balanced detection configuration. That is, if the same intensity of probe light falls on each diode, the net voltage produced is zero. This null condition applies in the absence of terahertz radiation. The presence of terahertz radiation will cause the detector pair to become unbalanced.

probe beam is in the crystal and at no other time. If the probe beam is of very short duration, the speed of electro-optic detection can be very fast, of a picosecond or less. This is very much faster than any thermal detector. If the probe signal is used to *gate* the detector – that is, to turn the detector on only at a specific time – the terahertz signal can be sampled on a very short time scale. This is the basis of time-domain spectroscopy. (The photoconductive detectors share the advantage of coherent detection.)

9.4 Photoconductive detectors

The **photoconductive detector** may be thought of as the inverse of the photoconductive emitter. A photoconductive detector is illustrated in Figure 9.7.

The conductivity of a piece of material, typically a semiconductor, is monitored. When terahertz radiation falls on the photoconductive material, the conductivity changes. The terahertz-frequency photons are usually of too low energy to create electron-hole pairs. Rather, they modulate the passage of a pre-existing current. This

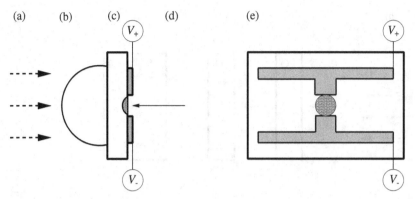

Figure 9.7 Photoconductive detector. (a) Pulsed terahertz radiation enters from the left (dashed arrows). (b) Optionally, a lens concentrates the terahertz radiation into the active region of the device. (c) The photoconductive substrate, transparent to terahertz radiation. An electrode structure is fabricated on the back of the photoconductive wafer. The structure is typically made of gold or another metal. The potential difference across the electrode pair is monitored. (d) Pulsed, synchronised, near-infrared or visible probe radiation enters from the right (full arrows). The probe radiation produces electron-hole pairs in the photoconductive slab, synchronised with the terahertz pulse. The movement of these charge carriers under the influence of the terahertz field is the origin of the potential difference measured across the electrodes. (e) Backside of the photoconductor showing a typical electrode structure. The probe beam illuminates the gap between the electrodes. The gap is typically 10 to 100 μm across.

current is generally produced by photoexcitation by a visible or near-infrared probe beam.

As with the electro-optic detection, the detection can be gated by the optical probe beam, permitting time-domain spectroscopy.

9.5 Summary

9.5.1 Key terms

sensitivity, 191 thermal detector, 195 electro-optic detector, 200
detectivity, 192 photoconductive detector, 201

9.5.2 Key equations

voltage responsivity $\quad R_V \equiv \dfrac{dV}{dP} \quad$ (9.1)

noise equivalent power $\quad \text{NEP} = \dfrac{V_{noise}}{R_V} \quad$ (9.2)

9.6 Table of symbols, Chapter 9

General mathematical symbols appear in Appendix B.

Symbol	Meaning	Unit
D	detectivity	W^{-1}
G	thermal conductance	$W\,K^{-1}$
NEP	noise equivalent power	W
NEP_{prb}	noise equivalent power per root bandwidth	$W\,Hz^{-1/2}$
P	radiant power	W
R_V	voltage responsivity	$V\,W^{-1}$
SNR	signal-to-noise ratio	[unitless]
T	temperature	K
T_0	temperature of reservoir	K
V	voltage output	V
V_{noise}	voltage output attributed to noise	V
Δf	bandwidth	Hz
$\tau_{thermal}$	thermal time constant	Hz^{-1}

Part III
Applications

10 SPECTROSCOPY

This chapter uses mathematics and the mathematics increases in sophistication as the chapter proceeds. Section 10.1 requires no mathematics. Section 10.2 assumes you know about sine and cosine and how to differentiate. Section 10.3 assumes you know about trigonometry and Fourier transforms. Section 10.4 assumes you can handle exponential notation and complex numbers.

Spectroscopy is the division of light into its separate frequencies. For example, sunlight is separated into its separate frequencies by raindrops when a rainbow forms. The separate frequencies make up the *spectrum*. The sun and the raindrops constitute a *spectrometer*, a device for producing a spectrum.

In the rainbow, the blue light is separated from the green light and the red light and so on. Breaking up light into its separate components is referred to as *analysis*. The opposite process is *synthesis* – making white light, for example, by combining red, green and blue, as is happening right now on the computer screen before me as I type.

Newton made a simple spectrometer by sending sunlight through a glass prism. The different colours emerged in different directions. This is the principle behind the *dispersive spectrometer*: the different colours are separated in space by an optical element such as a prism or a grating. The key characteristic of the core optical element is *dispersion*, that is, it acts on different frequencies differently. Section 10.2 deals with dispersive spectrometers.

An alternative way of separating frequencies depends on the Fourier theorem. As you already know from Chapter 2, Fourier analysis separates an arbitrary periodic function into its constituent frequencies. This is a purely mathematical concept. However, it can be realised in practice. One way is to split a beam of light into two parts and then bring the two parts together again. Bringing the light back together is referred to as *interference*; an instrument that does this is called an *interferometer*. The pattern of interference, the *interferogram*, is the Fourier transform of the spectrum. So calculating the Fourier transform yields the spectrum. The calculation is easily done with a computer. An instrument that operates like this is known as a *Fourier transform spectrometer*. As the technique was extensively developed and used by Michelson, the instrument is also known as a *Michelson interferometer*. We look into interferometry in Section 10.3.

A third way of doing spectroscopy is *time-domain spectroscopy*, the subject of Section 10.4. Time-domain spectroscopy has been developed only recently and is a key factor in the rapid rise in interest in the terahertz region in recent years. In time-domain spectroscopy, the same train of laser pulses is used both to excite the terahertz emitter and to trigger the terahertz detector. So in time-domain spectroscopy emitter and detector are synchronised and cannot be properly considered separately. The *time-domain* aspect comes into play when the detector trigger is delayed by different amounts of time relative to the emitter excitation. Mathematically speaking, the raw data are the same as from the interferometer: taking the Fourier transform of the time-domain spectrum yields the frequency-domain spectrum. Time-domain spectroscopy has several distinguishing features that set it apart from other forms of spectroscopy. First, it is usually implemented in pulsed mode, so short bursts or pulses of terahertz radiation are produced. The pulses often contain only a modest amount of energy, but because they are so short, the average power during the pulse can be very great. Moreover, since the detector is only gated when the excitation pulse is produced, the signal-to-noise ratio can be very high. Second, the mode of detection is sensitive directly to the electric field of the terahertz wave, whereas many other detection systems are sensitive only to the amplitude of the terahertz wave. The distinction here is that the electric field contains phase information, which is lost when only the intensity is measured. The distinction is sometimes expressed as being *coherent* as opposed to *incoherent* detection. The advantage is that both intensity and phase are obtained simultaneously. Put another way, the real and the imaginary parts of the wave are both determined at once.

So far we have thought about broadband sources of radiation and separating out the different frequency components. An alternative approach to spectroscopy is to start with narrowband or *monochromatic source* and somehow scan across a frequency range. This is the topic of Section 10.1.

What properties characterise a spectrometer? The first is the *spectral range*. This is the range of frequencies to which the spectrometer responds. Generally speaking, a large spectral range is desirable. For example, in the visible region, we would like a spectrometer that is sensitive to blue light, as well as to red and to green. The second is the *resolution*. This is the ability to distinguish (close) frequencies. For example, in the visible region, we would like a spectrometer that can distinguish yellow and orange light as falling between red and green. The third is *efficiency*. Generally speaking, if we have a longer time to make a measurement, we can expand the spectral range, or improve the resolution, or both. In practical applications, it is important to know what we can achieve in a given time.

Why are we interested in spectra? Largely because they tell us something about the material producing the spectrum. The colour of the sky tells something about the weather – 'red at night, shepherd's delight'. The colour of a banana can tell us whether it is under-ripe (green), ripe (yellow) or over-ripe (black). In this context, we often speak of the spectral *signature* of a material. For example, the terahertz spectrum of a colourless gas may reveal that it is water vapour and not carbon dioxide. The terahertz spectrum of a white powder may tell us if it is aspirin or cocaine.

Learning goals

After studying this chapter, you should be able to

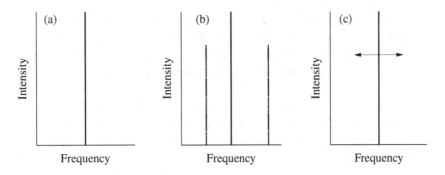

Figure 10.1 Monochromatic spectroscopy. (a) A single frequency source, such as a quantum cascade laser. (b) A set of discrete frequencies, such as from a molecular laser. (c) A single, adjustable frequency, such as from difference-frequency (beat) generation.

- explain the use of polychromatic and tuneable monochromatic sources for spectroscopy,
- understand the principles and limits to resolution of dispersive spectroscopy,
- outline the principle of operation of a Fourier-transform interferometer and explain the advantages over dispersive spectrometers,
- give an account of time-domain spectroscopy and obtain the optical constants from time-domain spectra.

10.1 Monochromatic sources

Spectra of monochromatic, or single-frequency, sources are shown in Figure 10.1a. An example is the quantum cascade laser. On the face of it, a monochromatic source can not be used for spectroscopy. Indeed, according to our definition, a spectrum requires a range of frequencies. Sometimes monochromatic sources are used for 'spectroscopy' when another variable (than frequency) is varied. For example, the change in the sample transmission at the single frequency might be measured as the magnetic field is varied and the result described as *magnetospectroscopy*. Or the sample transmission might be measured as a function of temperature, or of time. Interesting as these experiments are, they are not spectroscopy in the sense we are using the word here.

Figure 10.1b shows a source with many different, separated frequencies available, but with only one being emitted at a given time. This might be called a poly-monochromatic source, or a **polychromatic source**, so long as it is clear that not all frequencies in the range are available, but only a limited number of distinct frequencies. The molecular gas laser is an example of this type of source. It is tuned to the desired laser transition by varying parameters such as the pump line, the gas, the pressure and the cavity length. (Sometimes more than one laser line is excited, but usually this is avoided.)

Spectroscopy can be carried out using such a source, but it is relatively time-consuming and laborious to change between the different frequencies. The source is tuneable, but only to discrete frequencies, and these are not necessarily evenly spaced. We may describe this source as **discontinuously tuneable**, in contrast to the source depicted in Figure 10.1c.

Figure 10.1c represents a **continuously tuneable** source. At any given time, the source is emitting a single frequency. But the frequency may be controlled in some way. Examples are backward-wave oscillator and two-colour (beat-frequency difference) sources.

A practical advantage of tuneable spectrometers is that their resolution may be very high. The resolution is determined on the one hand by how narrow the source emission is and on the other hand by how accurately it can be tuned. (There is no point having fine tuning if the source is intrinsically broad, nor to having a narrow source if the tuning cannot be controlled.) Another advantage is that the spectrometer need only cover the frequency range of interest. It shares this advantage with the dispersive spectrometer; in contrast, the interferometer and the time-domain spectrometer collect all frequencies (that they are able to) simultaneously. The ability to restrict the frequency range to the one of interest may lead to greater efficiency, as all the experimental time is dedicated to relevant measurements. The main disadvantages in practice are that the range of frequencies over which the source may be tuned is usually limited and that scanning over a wide frequency range may take a long time.

The sources depicted in Figure 10.1 all have sharp (monochromatic) characteristics. They are usually based on lasers, as in the examples I have given, the quantum cascade laser, the molecular gas laser and two mixed near-infrared lasers. An approximation to a monochromatic source can also be made by starting with a broadband source and removing unwanted frequencies with a band-pass filter. At the dawn of this field, reststrahlen reflection filters were used for this purpose. After multiple reflections, most intensity was removed from the beam, except in the band between the LO and TO phonons. This frequency range is relatively large (of the order of a terahertz) and tuning the frequency, by swapping in different crystals, time-consuming; moreover, the whole apparatus is cumbersome. Further, the method is inefficient in that much of the radiation from the source is thrown away to produce the quasi-monochromatic output. The method of spectroscopy by filtering a broadband source is not widely used today.

10.2 Dispersive spectroscopy

In conventional spectrometers, the three main components are the source, the dispersing element and the detector. In the historical example of Newton, the source was the sun, the dispersing element a glass prism and the detector Newton's own eyes, looking at the spread of colours cast on a viewing screen. This is an example of **dispersive spectroscopy**.

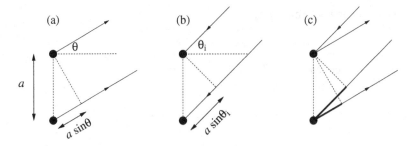

Figure 10.2 Diffraction grating geometry. The grating comprises N elements with spacing a. Two elements are shown here. (a) Parallel rays leaving the grating. The bottom ray is longer by $a\sin\theta$. (b) Parallel rays arriving at the grating. The bottom ray is longer by $a\sin\theta_i$. (c) In total, for incidence angle θ_i and exit angle θ, the bottom path exceeds the top path by $a(\sin\theta + \sin\theta_i)$. The total path difference is indicated by the bold segments.

Some spectroscopy is interested in the spectral characteristics of the source; in Newton's case, the sun. This is **emission spectroscopy** – the object of study is the emitter. Only the dispersive element and the detector are required in an emission spectrometer.

Other spectroscopy is concerned with the characteristics of a sample, how it transmits or reflects light of different frequencies. In this case, the spectrometer source must be matched to the sample frequencies we wish to measure. For example, to study the ultra-violet blocking properties of a sunscreen, a source that emits in the ultra-violet is needed. In terahertz spectrometers, we use a source such as one described in Chapter 7.

In the visible region, a prism, of glass or other suitable material, is often used as the dispersing element. No suitable material has yet been found with the requisite properties – high dispersion and high transmission – to make a good dispersive prism in the terahertz region.

In the visible region, as an alternative to a prism, a **grating** may be used as the dispersive element. A grating for the dispersive element is also feasible in the terahertz region. The grating has alternating reflecting and non-reflecting strips. In the terahertz region, this is achieved relatively easily by using wires, wound around a former. The light transmitted (or reflected) by the grating at different angles is the sum of the transmission (or reflection) from all the different segments. The various contributions interfere, adding or subtracting, resulting in a diffraction pattern. For this reason, such a grating is known as **diffraction grating**.

The geometry of the diffraction grating is shown in Figure 10.2. The diffraction grating consists of a line of objects separated by the distance a. Let us look at the path differences involved in rays arriving from or leaving at arbitrary directions. The total path difference between a ray that arrives at incident angle θ_i and leaves at angle θ is $a(\sin\theta + \sin\theta_i)$, as demonstrated in Figure 10.2. The subscript 'i' in θ_i refers to incident angle, which we take as fixed; we allow the departure angle θ to vary. Now, if the path difference happens to correspond to an integral number of wavelengths, reinforcement will occur. For the integer m ($= 0, \pm 1, \pm 2, \ldots$) the corresponding angle of departure is denoted θ_m and determined by

$$m\lambda = a(\sin\theta_m + \sin\theta_i). \tag{10.1}$$

In terms of frequency $f = c/\lambda$, this may be written

$$mc = fa(\sin\theta_m + \sin\theta_i). \tag{10.2}$$

This relationship is known as the *grating equation*.

Let's look at three special cases.

$\theta_i = 0$ This condition corresponds to zero angle of incidence, in other words, normal incidence. The grating equation then becomes

$$mc = fa\sin\theta_m. \tag{10.3}$$

$\theta_i = \theta_m$ This condition corresponds to the light leaving along the same path it arrived by. There is the same path difference on arrival as on departure; the total path difference is twice this. The grating equation then becomes

$$mc = 2fa\sin\theta_m. \tag{10.4}$$

$\theta_i = -\theta_m$ This condition corresponds to the light arriving and leaving at the same angle, but on opposite sides of the grating normal. You might associate this with the usual condition for reflection, where the angle of incidence equals the angle of reflection. One ray has an additional path length on arrival, the other on departure, the additional path lengths are the same, the net effect being the total path lengths are the same. In other words, there is no path difference. The grating equation then becomes

$$m = 0. \tag{10.5}$$

The meaning of this is that the only order available is the zeroth order; all frequencies are scattered in the same direction; the diffraction grating is not exhibiting diffraction as such, merely reflection.

The last two of the special cases illustrate that we need to be careful with the signs of the angles. As shown in Figure 10.2, angles are measured in the usual way, to be positive if they are in an anticlockwise sense from the reference line, here the grating normal. Both θ_i and θ_m are positive in the figure. If the angles were to the other side of the normal (corresponding to the rays coming from below rather than coming from above in the figure), they would be negative.

Likewise, we need to be careful about the sign of the path difference. In Figure 10.2, I have denoted the amount by which the bottom ray exceeds the top ray the path difference. This is positive for the rays shown in Figure 10.2. If we choose the opposite possibility, measuring the amount by which the top ray is greater than the bottom ray, the sign of the path difference would reverse.

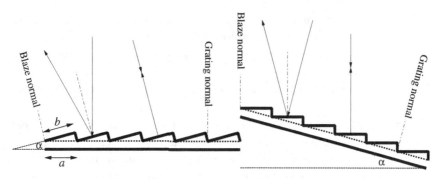

Figure 10.3 A blazed grating. The period of the grating is a. The blaze angle is α. The blaze angle is usually small; here it is $\pi/12$. The length along each ramp is $b = a\cos\alpha$. A ray parallel to the grating normal strikes the ramp at an angle of incidence equal to the blaze angle. It is deviated through a total of twice the blaze angle. This is shown as the left ray in each part of the figure. A ray parallel to the blaze nomal strikes the ramp at normal incidence. It is reflected along the same path it came. This is shown as the right ray in each part of the figure.

A diffraction grating might be made along the lines indicated in Figure 10.2 by a grid of metal wires, spaced by the distance a. You might imagine two of the wires running into the page through the dots in Figure 10.2. Although this will work, it is inefficient. Much of the light is simply reflected. This light appears in the **zeroth order** ($m = 0$) and there is no separation of frequencies. At the **higher order** ($m = \pm1, \pm2, \ldots$) angles, where there is a separation of frequencies, the light intensity is relatively weak.

A cunning way to make an efficient diffraction grating is to use an **echelon** geometry. The point of this is to shift energy from the zeroth order, where the frequencies are not separated, to a higher order, where they are. First I will describe an echelon or **blazed grating**, then explain how it works.

A blazed grating is shown in Figure 10.3. It consists of a set of identical ramps. The angle of ramp inclination is called the **blaze angle**. In the figure the blaze angle is $\alpha = \pi/12$; in practice it is often smaller. The period of the grating is a. The inclined length of each ramp is $b = a\cos\alpha$.

To explain how it works, let's consider two special rays striking the blazed grating.

Normal incidence A ray striking the blazed grating at normal incidence will usually encounter a flat surface inclined at angle α to it. (Sometimes the ray may strike the side of the ramp, rather than the top. This becomes less likely as α becomes smaller.) The ray is then reflected. The angle of reflection is equal to the angle of incidence so the ray is deflected through twice the blaze angle.

Where is the zeroth order found? We solve the grating equation with $\theta_i = 0$ and $m = 0$ to find

$$0 = fa(\sin\theta_0 + \sin(0)); \qquad (10.6)$$

so the angle of the zeroth order diffraction is $\theta_0 = 0$; in other words, the zeroth order diffraction is produced directly behind the incoming ray. This is the same as for the standard (unblazed) grating; what is different here is that most of the incoming ray intensity does not end up in this direction, but somewhere altogether different: deflected through twice the blaze angle.

If we could arrange for this strong beam, deflected through twice the blaze angle, to match up with a higher order ($m = \pm 1$ or $\pm 2\ldots$), we might be onto something. In fact, this is exactly what happens in the *autocollimation* geometry.

Autocollimation The term **autocollimation** refers to the incident beam being reflected back along the path it came. For a plane mirror, or for the usual diffraction grating, this condition is fulfilled for normal incidence. As we have already seen in Figure 10.3, normal incidence on the blazed grating results in a beam deflected through twice the blaze angle. We need to set the incidence angle to the blaze angle to obtain a reflected beam along the original path for a blazed grating; this is also shown in Figure 10.3.

Blazed gratings are usually used in the autocollimation geometry. That is, the grating is inclined relative to the incoming radiation at the blaze angle.

Where is the zeroth order found in this case? We solve the grating equation with $\theta_i = \alpha$ and $m = 0$ to find

$$0 = fa(\sin\theta_0 + \sin(\alpha)); \tag{10.7}$$

so the angle of the zeroth order diffraction is $\theta_0 = -\alpha$. This means the zeroth order is found on the far side of the grating normal to the incoming beam.

The point of the blazing is to bring the strong, specularly reflected beam into coincidence with one of the non-zero orders of diffraction. In other words, we set θ_m to be equal to the angle of incidence, which is already set at the blaze angle. Then the grating equation reads

$$mc = 2fa\sin\alpha. \tag{10.8}$$

Once the order is chosen (usually a small order is chosen, $m = 1$ or 2) and the blaze angle decided (which is normally small, but depends on the exact geometry of the instrument), the appropriate grating period may be calculated for the desired operating frequency.

Of course, the role of a spectrometer is to operate over a range of frequencies, not just one, but these calculations give the optimum arrangement for one operating frequency.

Angular dispersion

The **angular dispersion** relates the change in angle of the diffracted beam to the change in wavelength (or frequency) of the incident beam.

We begin with the wavelength version of the grating equation.

$$m\lambda = a(\sin\theta_m + \sin\theta_i). \tag{10.9}$$

We will assume that θ_i is fixed; that is, we will not tilt the grating relative to the incoming beam. We will allow the angle of the diffracted beam to vary and so drop the m subscript on θ_m, but retain the m related to the order, since we are not changing λ or θ so much as to move to another order. Differentiating with respect to θ gives

$$m\frac{d\lambda}{d\theta} = a\cos\theta, \qquad (10.10)$$

or

$$\frac{d\lambda}{d\theta} = \frac{a\cos\theta}{m}, \qquad (10.11)$$

or, on inversion,

$$\frac{d\theta}{d\lambda} = \frac{m}{a\cos\theta}. \qquad (10.12)$$

In terms of frequency $f = c/\lambda$, this may be written

$$\frac{d\theta}{df} = \frac{-mc}{af^2\cos\theta}. \qquad (10.13)$$

Line broadening
A monochromatic source produces an extended diffraction pattern. I won't prove here but will simply state that the half-width of the principal maximum is

$$\Delta\theta = \frac{\lambda}{Na\cos\theta} = \frac{c}{Naf\cos\theta}. \qquad (10.14)$$

Here N denotes the number of periods in the grating and so Na is the width of the grating.

Chromatic resolving power
The **chromatic resolving power** is defined as

$$R \equiv \frac{\lambda}{\Delta\lambda}. \qquad (10.15)$$

Starting with Equation (10.12) and replacing $d\theta$ with the expression for $\Delta\theta$ in Equation (10.14), and replacing $d\lambda$ with $\Delta\lambda$ allows Equation (10.15) to be written as

$$R = Nm. \qquad (10.16)$$

So the chromatic resolving power depends directly on the number of periods of the grating, N, as well as the order number, m.

Eliminating m from the grating equation, we can write

$$R = Na\frac{\sin\theta_m + \sin\theta_i}{\lambda} = Naf\frac{\sin\theta_m + \sin\theta_i}{c}. \qquad (10.17)$$

The resolving power is maximised when $\sin\theta_m = 1$ and $\sin\theta_i = 1$, that is, when the corresponding angles are $\pi/2$. This corresponds to grazing incidence which, of course, is impractical.

Exercise 10.1 Chromatic resolving power
Determine the chromatic resolving power of a diffraction spectrometer furnished with a grating ruled at 100 lines per millimetre and 10 cm wide. ∎

Higher orders

A complication of the grating spectrometer is that higher orders also appear. For example, the angular position relative to the grating where the first order of 1 THz radiation appears also corresponds to the point where the second order of 2 THz radiation appears. There are two immediate consequences. First, if the interest is in 1 THz radiation and not 2 THz radiation, suitable filtering will be needed to remove the unwanted radiation from the spectrometer. Second, the useable range of the grating spectrometer is limited to one octave; in this example, from 1 to 2 THz. If a greater range is needed, several gratings are usually employed. Of course, there is a cost in time to change between gratings.

Detector

For the detector, many of the detectors described in Chapter 9 may be used.

Scanning

The spectrum is collected by measuring the angular distribution of light from the grating. In principle, this can be accomplished by moving either the source, the grating or the detector. In practice, the source and detector are usually fixed and the grating is rotated. So the source to grating, and grating to detector angle are both changed, and the frequency of light now falling on the detector has changed. Another way to collect the light intensity at a different angle would be to employ several detectors, or an array of detectors. This is common for visible-light spectrometers, where charge-coupled device (CCD) detectors are available with, say, 1024 elements. Such multiple-element detectors are not routinely available in the terahertz region, although work is under way to develop them.

10.3 Interferometry

The scheme of the Michelson interferometer is shown in Figure 10.4. The basic idea of **interferometry** is that the light is split – by a device called a beamsplitter – and later the two beams are recombined, or *interfere* with each other. Typical beamsplitters in the terahertz region are made by sheets of plastic or semiconductor wafers. Some of the light is transmitted and some of the light is reflected. Ideally, half the light is transmitted and half reflected, but it is difficult in practice to accomplish this over a wide frequency range. Usually, the beamsplitter is set at 45° to the incoming light. In this geometry, the transmitted light goes essentially straight through the beamsplitter (in fact, the transmitted beam is parallel to but offset slightly from the incoming beam because of refraction at the front and back surfaces of the beamsplitter) and the reflected light is deflected through a right angle.

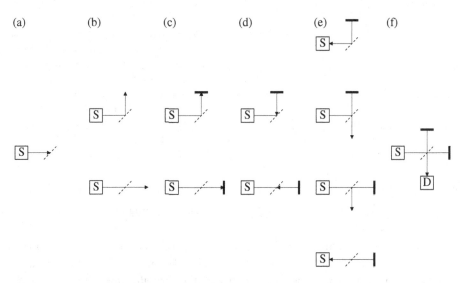

Figure 10.4 Beam paths in the Michelson interferometer. S denotes source and D denotes detector. (a) The beam emitted from the source is directed to a beamsplitter. At the beamsplitter, (b), part of the beam is reflected, and part transmitted. (Ideally, half the beam will be transmitted, half reflected.) The two portions are directed to two different mirrors, (c), and reflected from them, (d). Back at the beamsplitter, the beams are again either reflected or transmitted, (e). Thus there are four possible beam paths. Two of these are combined and detected – the beam that was first reflected then transmitted at the beamsplitter and the beam that was first transmitted then reflected at the beamsplitter, (f).

The two beams of light then proceed to two flat mirrors which they strike at normal incidence. They return along their original paths. We now concentrate on the originally transmitted beam and the portion of that beam that is reflected at the back of the beamsplitter. (Some of the beam coming back from the mirror is transmitted by the beamsplitter, but we will ignore that.) The originally reflected beam also comes back to the beamsplitter. We will concentrate on the portion of that now transmitted by the beamsplitter (not the portion reflected by the beamsplitter we will ignore this). These are the two beams that interfere, and the combination of the two is measured at the detector.

So far, so good. We have seen two beams are separated by the beamsplitter and re-combine at the detector. The interferometer can be set up so that the two beam paths are of exactly the same length. In other words, there is no difference in the path length. This configuration is shown in Figure 10.5a. When set at **zero path difference**, the light in one beam will be in phase with the light in the other beam. When recombined at the detector, the two beams will constructively interfere. We will retrieve the original signal.

How is the Michelson interferometer used for spectroscopy? By moving one of the mirrors and so changing the path length of one of the beams. Then, when the beams recombine, they will be at different phases. The phase change depends on how the change in path length relates to the frequency of the light. The interference may now be constructive, or destructive, or something in between. Analysing the pattern of

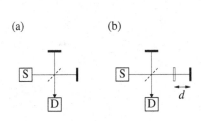

Figure 10.5 The Michelson interferometer, comprising source (S), detector (D), beamsplitter and two mirrors. (a) The path length from the source to the detector is the same via either mirror. This configuration is zero path difference. (b) Now one mirror has moved a distance d. The light to this mirror has to travel an additional d in getting there and d in getting back. The increase in optical path length is $2d$.

interference with path difference – the interferogram – allows us to retrieve the spectrum.

Figure 10.5b shows the Michelson interferometer with one of the mirrors moved a distance d from the position of zero path difference. The light now has to travel an additional distance d to get to the mirror and then another amount d to return to the original position of the mirror, so the path length is increased by $2d$. (We have assumed the interferometer is in an environment with refractive index 1, which is a good approximation for air; if the path is in a medium of refractive index n, then the **optical path difference** as the mirror moves a distance d is $2nd$.)

In analysing the operation of the interferometer, let us start with the simple situation of light of a single frequency. If the two separate paths, going via the two different mirrors, are exactly the same length, the two beams will arrive at the detector at the same time, or in phase, and so add together, or constructively interfere. (To be more precise, it is not the physical path length that must be the same, but the optical path length, taking into account the refractive index of the materials through which the light is travelling.) Now, let one mirror be moved by one quarter of the wavelength of light. The light striking that mirror will have to travel a total of one half a wavelength farther than before (one quarter before striking the mirror, one quarter after striking the mirror). Compared to the other beam, it will now be out of phase by half a wavelength, in other words, it will be exactly out of phase or antiphase to the other light beam. The two will cancel, or completely disappear. Nothing will be recorded at the detector. So, as the position of one mirror is scanned, the signal detected will fluctuate from zero to a maximum to zero again regularly. To obtain the frequency of the source requires nothing more than to measure the optical path between successive maxima; this gives the wavelength.

To consider the next most simple situation, let us consider a duo-chromatic source emitting two different frequencies at the same intensity. At zero path difference, as always, we will have constructive interference, and the full source intensity will be reproduced. As one mirror now scans, it gives the same path difference to each of the two frequencies, but this corresponds to a different phase difference. The effect will be of adding two oscillations at slightly different frequencies; a beat pattern results.

10.3 Interferometry

In general, as will be shown in Section 10.3.3, the interferogram is the Fourier transform of the spectrum. Taking the inverse Fourier transform of the interferogram then yields the spectrum. As will be shown in Section 10.3.4, the resolution of the spectrum improves as the scan length increases.

10.3.1 Slow scanning

A **step-scan interferometer** dwells at one point until data of sufficient signal to noise has been acquired, then steps to the next point. Typically, a lock-in amplifier is used in detecting the signal, and a sufficient dwell time is set at each point for at least several time constants of the lock-in amplifier to go by. If the Fourier transform of the interferogram is calculated at each step (in 'real time', which is quite feasible by computer) then the resulting spectrum exhibits little noise, and the resolution increases as the experiment continues. The decision can be made to terminate the scan when sufficient resolution has been attained. Typically, the scan takes several minutes to several tens of minutes to complete. The advantage in this method is that the experiment can be continued until the resolution is at the desired level; a disadvantage is that the signal-to-noise ratio is set at the start and cannot be increased later.

10.3.2 Rapid scanning

A **rapid-scan interferometer** quickly scans over the whole optical delay length, repeatedly. If the Fourier transform of the data is calculated after the first scan, the resulting spectrum will be of the chosen resolution, but the spectrum will be rather noisy. As the second scan is taken, and added to the original scan, the noise tends to average out, and a less noisy spectrum results. The decision can be made to terminate the scanning when an acceptable level of signal-to-noise ratio is obtained. Typically, the scans take a few seconds or tens of seconds to complete, and tens to hundreds of scans are added together. The advantage in this method is that the experiment can be continued until the signal-to-noise ratio is at the desired level; a disadvantage is that the resolution is set at the start and cannot be increased later.

10.3.3 Theory

Following the development of Chapter 3 in combining oscillations, two electric fields of the same frequency f and angular wavenumber k and electric field amplitudes E_1 and E_2 combine to give an electric field of amplitude E_0 given by

$$E_0^2 = E_1^2 + E_2^2 + 2E_1 E_2 \cos \delta, \qquad (10.18)$$

where δ is the phase difference between them. In terms of the intensity,

$$I_0 = I_1 + I_2 + 2\sqrt{I_1 I_2} \cos \delta. \qquad (10.19)$$

For simplicity, let us assume the intensities of the two fields are equal, $I_1 = I_2$, then

$$I_0 = 2I_1 + 2\sqrt{I_1^2} \cos \delta = 2I_1(1 + \cos \delta). \qquad (10.20)$$

With slight change of notation we can write

$$I = \frac{I_0}{2}(1 + \cos\delta). \tag{10.21}$$

As the path difference changes, I fluctuates between a maximum of I_0 (at $\delta = 0, \pm 2\pi, \pm 4\pi \ldots$) and a minimum of zero (at $\delta = \pm\pi, \pm 3\pi \ldots$).

The mathematical result is profound. Into the interferometer goes a monochromatic wave, characterised by frequency f. Out comes a signal which, when plotted as a function of phase, is a single cosine. In effect, the interferometer has acted as an analogue computer, taking a single frequency and producing the Fourier transform, a cosine.

Normally, we record the experimental data as a function of mirror travel, which we previously denoted d but will call here x, measured relative to the position of zero path difference. (Usually the mirror moves at a constant speed, and we could record it as a function of time t.) So, at a particular mirror position,

$$I(x) = \frac{I_0}{2}(1 + \cos(kx)) = \frac{I_0}{2}(1 + \cos(2\pi\sigma x)). \tag{10.22}$$

If we integrate over the whole of wavenumber space at any x we get

$$I = \int_{-\infty}^{+\infty} I(\sigma)d\sigma = \int_{-\infty}^{+\infty} \frac{I_0}{2}(1 + \cos(2\pi\sigma x))d\sigma. \tag{10.23}$$

Now, let us suppose the source is not monochromatic, but has some spectral distribution we will write as $B(\sigma)$, keeping σ as the variable. Then the intensity will be given by a whole collection of equations like the preceding one, weighted by B. Thus

$$I(x) = \int_0^\infty B(\sigma)[1 + \cos(2\pi\sigma x)]d\sigma. \tag{10.24}$$

This may be separated into two parts – the intensity at zero path difference, and the rest:

$$\begin{aligned} I(x) &= \int_0^\infty B(\sigma)d\sigma + \int_0^\infty B(\sigma)\cos(2\pi\sigma x)d\sigma \\ &= \frac{I(0)}{2} + \int_0^\infty B(\sigma)\cos(2\pi\sigma x)d\sigma. \end{aligned} \tag{10.25}$$

Taking the Fourier transform, and ignoring the zero-path intensity,

$$B(\sigma) = \int_0^\infty I(x)\cos(2\pi\sigma x)dx. \tag{10.26}$$

So, there it is. The Fourier transform of the interferogram gives the spectrum.

10.3.4 Resolution

The farther the mirror travels, the higher the resolution of the spectrum. The resolution of the Michelson interferometer may be estimated using the principle behind the Rayleigh criterion: namely, when the maximum of one frequency overlaps with a minimum of an adjacent frequency in the observed pattern. In this case, the pattern is the interferogram; the controlling variable is not the angle but rather the optical path difference.

It is instructive to consider the resolution in terms of wavelength, in terms of wavenumber and in terms of frequency, so I will discuss each. It is also instructive to consider relative resolution and absolute resolution, so I will discuss each of those as well.

Imagine we have two monochromatic sources of the same intensity but of slightly different frequencies f_1 and f_2, with $f_1 < f_2$. The corresponding wavenumbers will be $\sigma_1 = f_1/c < \sigma_2 = f_2/c$. The corresponding wavelengths will be $\lambda_1 = c/f_1 > \lambda_2 = c/f_2$.

Let's see what happens as the interferogram is built up as the optical path length d increases. At zero path difference the two sources will be in phase and the intensity twice that of either source. As the path length difference increases, the two sources will gradually come out of phase, the next maximum in the signal from source 1 appearing at $d = \lambda_1$, slightly after the maximum from source 2 appearing at $d = \lambda_2$. As we increase the path length farther, the phase difference between the two sources increases. If we go far enough, the two sources will be completely out of phase.

The first time a maximum of one source coincides with a minimum of the other source is when the shorter-wavelength source has gone through an additional half-cycle. If we call the number of cycles in common m, this condition is written

$$m\lambda_1 = (m + 1/2)\lambda_2 = d_m. \tag{10.27}$$

The closer the frequencies are to begin with, the farther we will have to go, the larger m will have to be, the larger the corresponding path length d_m will have to be, before this condition is met. The condition may be written in terms of frequency and angular wavenumber as

$$m\frac{c}{f_1} = (m + 1/2)\frac{c}{f_2} = d_m, \tag{10.28}$$

and

$$\frac{m}{k_1} = \frac{m + 1/2}{k_2} = \frac{d_m}{2\pi}. \tag{10.29}$$

We may rearrange Equation (10.27) to obtain

$$\lambda_1 - \lambda_2 = \frac{\lambda_2}{2m} = \frac{\lambda_1 \lambda_2}{2d_m}. \tag{10.30}$$

This is the smallest wavelength difference that is resolved. We may express this in terms of a quality factor of the wavelength divided by the resolvable wavelength difference,

$$\frac{\lambda_2}{\lambda_1 - \lambda_2} = 2m = \frac{2d_m}{\lambda_1}. \tag{10.31}$$

The *resolving power*, written in this way, is the same as the resolving power of a diffraction grating, Nm, with N, the number of sources, here being two, corresponding to the two mirrors.

In terms of frequency, starting from Equation (10.28), the minimum resolvable frequency difference is

$$f_2 - f_1 = \frac{f_1}{2m} = \frac{c}{2d_m} \qquad (10.32)$$

and the resolving power

$$\frac{f_1}{f_2 - f_1} = 2m = \frac{2f_1 d_m}{c}. \qquad (10.33)$$

Likewise, in terms of angular wavenumber,

$$k_2 - k_1 = \frac{k_1}{2m} = \frac{\pi}{d_m} \qquad (10.34)$$

and

$$\frac{k_1}{k_2 - k_1} = 2m = \frac{2k_1 d_m}{\pi}. \qquad (10.35)$$

The resolving power is the same, $2m$, whether calculated using wavelength, frequency or angular wavenumber. In contrast, the equations for absolute resolution for frequency and angular wavenumber only involve the scan length d_m and not frequency or angular wavenumber in the rightmost expressions. In contrast, the absolute wavelength resolution involves a term in wavelength squared (Equation 10.30). Equation (10.34) is especially convenient: the resolution, in wavenumbers, is simply the inverse of the scan length (with a factor of two).

Exercise 10.2 Resolution

Two monochromatic sources of frequency close to 1 THz are monitored by an interferometer. The 2000th fringe of one source coincides with the 2001st fringe of the second. What is the resolving power? If the first source is at a frequency of exactly 1 THz, what is the frequency of the second source? ■

10.3.5 Advantages of interferometers over dispersive spectrometers

Fourier transform interferometers have two main advantages over dispersive instruments.

Throughput, or Jacquinot, advantage

The amount of light reaching the detector in the dispersive spectrometer is limited by the slits. In contrast, a wide beam of light is employed in the interferometer. The amount of light getting through to the detector is much greater. Although the resolution of the interferometer limits the beam radius, the beam area is still much larger than for a grating instrument of the same resolution. Loosely speaking, the dispersive spectrometer employs a one-dimensional beam, the interferometer a two-dimensional one.

Multiplex, or Fellgett, advantage

The dispersive spectrometer collects data for light of one frequency at a time. Then it moves on, to the next frequency. The interferometer collects data from all frequencies at once.

10.4 Time-domain spectroscopy

10.4.1 Time-domain spectroscopy and interferometry

Many aspects of **time-domain spectroscopy** are shared with interferometry. There are two beams of light, originating from the same source (in this case the excitation laser), that are split. One beam goes to the emitter. The other goes to the detector. One of the beams (typically the one to the detector) is delayed relative to the other. The same mathematics applies as with the interferometer. The raw data of detector signal is given as a function of the delay, there called the interferogram, here called the time-domain spectrum. The inverse Fourier transform of the data is calculated, yielding the spectrum in the frequency domain.

The same considerations regarding resolution and bandwidth that applied for the interferometer apply to time-domain spectroscopy. The resolution increases as the time delay increases. The bandwidth increases as the density of data points increases.

10.4.2 Time-domain spectroscopy configurations

Both the emitter and the detector may be electro-optical crystals. In this case no external electrical bias need be applied. This is a purely optical configuration. On the other hand, photoconductive antennas and detectors may both be employed. Alternatively, a photoconductive emitter and electro-optic detector, or, less commonly, an electro-optic emitter and photoconductive detector may be used.

10.4.3 Extracting optical constants – overview

A key feature of time-domain spectroscopy is that it allows the user to measure the amplitude and the phase of the electromagnetic wave at the same time. This means it is possible to obtain the real and imaginary parts of the optical constants simultaneously. In this section, I explain in general terms how this is done. In Section 10.4.5 I give a more detailed mathematical treatment.

First, a reference spectrum is collected without the sample in the beam path. The reference beam path need not be empty, in other words, be a vacuum (although sometimes it is). The main requirement is that the reference path does not change when the sample is inserted. For example, if the reference path is filled with air at a certain humidity, the humidity should not change after the sample is inserted or a poor analysis will result. For simplicity here, let us assume that the reference path is indeed a vacuum. After the reference spectrum is collected, the sample is inserted in the beam path. A sample spectrum is then collected. The two spectra are compared to yield absolute information about the sample.

We may divide the optical constants into two parts, real and imaginary. The real part is called the refractive index. The imaginary part is called the absorption coefficient. The real part is associated with the change in the speed of light as light goes from a vacuum into the material. The imaginary part is associated with the absorption or loss of energy from the light into the sample.

Let's consider in general terms how the time-domain spectrum is affected by changes in the optical constants. If we replace a section of vacuum (refractive index 1) with a sample of another refractive index, this will introduce a time delay to light travelling along the path. Let's say the thickness of the sample is d. Originally, the time taken for light to travel the distance d in the vacuum was

$$t_v = \frac{d}{c}. \tag{10.36}$$

Through the sample, light travels at the reduced speed c/n. So with the sample in place the time to traverse the same distance is

$$t_s = \frac{nd}{c}. \tag{10.37}$$

Hence the time delay introduced by a sample of thickness d and refractive index n is

$$t_s - t_v = \frac{nd}{c} - \frac{d}{c} = (n-1)\frac{d}{c}. \tag{10.38}$$

Example 10.1   Silicon wafer

A wafer of silicon (refractive index 3.5), 350 μm thick, is introduced into a light path. By how much is the light retarded?

Solution
We use Equation (10.38) with $d = 350$ m^{-6}, $n = 3.5$ and $c = 299\,792\,458$ m/s to obtain $\Delta t = 3$ ps.

Exercise 10.3   Optical delay

A 1 mm thick piece of polyethylene ($n = 1.2$) is inserted into one arm of a spectrometer. What is the time delay so introduced? ■

The effect of the refractive index in delaying the light pulse is not surprising: the light travels more slowly in the material than in the vacuum, and this change in speed is precisely what the refractive index signifies. There is another, more subtle, effect of the refractive index: to reduce the magnitude of the light pulse as it encounters the sample. Let me point out, the refractive index does not cause the sample to absorb radiation; the absorption coefficient signifies how much absorption occurs. Rather, the mismatch of refractive index at the interface between the vacuum and the sample causes a reflection of light at the front surface. Less light comes through the sample, not because the sample is absorbing some, but because the sample is sending some light back to where it came from.

We see now that the effect of a non-unity refractive index on the original light pulse will be twofold: the pulse will be both delayed and reduced, relative to the reference pulse. It is a relatively simple matter to reverse the process, and, by comparing the sample and reference spectra, deduce the refractive index.

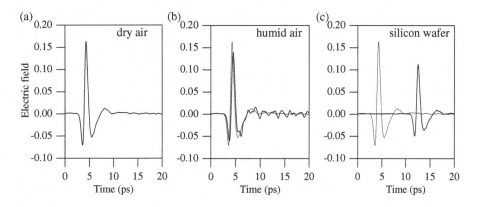

Figure 10.6 Time-domain spectroscopy of (a) dry air, (b) humid air and (c) a silicon wafer.

What of the role of the absorption coefficient? The most obvious effect is to reduce the magnitude of the light pulse. This reduction in magnitude can be distinguished from the reduction in magnitude connected with refractive index as it is not accompanied by a temporal shift. So, even given non-zero absorption coefficients and refractive indices, it is a relatively simple matter to analyse the spectrum to extract both. Section 10.4.5 deals with this in more detail.

10.4.4 Examples

Figure 10.6 gives three examples of time-domain spectra.

Figure 10.6a shows the time-domain spectrum of dry air. The main feature is a large pulse. The maximum of the pulse corresponds to the probe beam arriving in the detector at the same time as the terahertz radiation excited by the pump beam does. The pulse is about 1 ps wide; this means the spectral bandwidth is about 1 THz.

Figure 10.6b shows the time-domain spectrum of humid air. (The spectrum of dry air, Figure 10.6a, is included for comparison.) The height of the main pulse is slightly reduced and the main pulse slightly shifted in time for the humid air relative to the dry air. The shift is due to the change in refractive index. The reduction in height is partly due to the change in refractive index, partly due to the absorption in the humid air. The most obvious difference between the spectra is the oscillations after the main pulse in the humid air. These oscillations are not noise; they are the signature of the spectrum of the water vapour.

Figure 10.6c shows the time-domain spectrum of a silicon wafer. (The spectrum of dry air, Figure 10.6a, is included for comparison.) The most obvious difference between the spectra is that the prominent pulse has been shifted in time. It has also been reduced in height. These differences are mainly due to the larger refractive index of silicon than dry air. There is little absorption in the silicon; the change of height is largely due to light being reflected.

226 SPECTROSCOPY

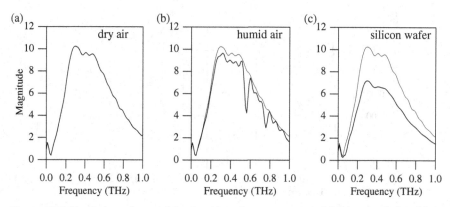

Figure 10.7 Fourier transforms of the time-domain spectroscopy of (a) dry air, (b) humid air and (c) a silicon wafer. The Fourier transform results in a complex quantity; here the magnitude of the complex quantities is shown.

Figure 10.7 gives the Fourier transforms of the three time-domain spectra of Figure 10.6.

Figure 10.7a shows the frequency-domain spectrum of dry air as calculated from the time-domain spectrum. The overall shape – a steep rise and fall – gives the instrument function. That is, the spectrometer is limited at how low and how high a frequency it can operate, and the rise and fall in the spectrum reflects this rather than the properties of the reference material, dry air. (If a vacuum, rather than dry air, was in the spectrometer, a similar spectrum would result.)

Figure 10.7b shows the frequency-domain spectrum of humid air as calculated from the time-domain spectrum. (The spectrum of dry air, Figure 10.7a, is included for comparison.) The spectrum for humid air closely tracks the spectrum for dry air, except around particular frequencies, such as 0.5 THz, where sharp dips are seen. These sharp dips are characteristic resonances of water vapour.

Figure 10.7c shows the frequency-domain spectrum of a silicon wafer as calculated from the time-domain spectrum. (The spectrum of dry air, Figure 10.7a, is included for comparison.) The spectrum for silicon is similar in shape to the spectrum for dry air, except reduced in height. There are no characteristic resonances, but an overall reduction in signal. It might be guessed that the reduction in signal is due to absorption in the sample but, as I have mentioned previously, in this case it is due to light not passing through the sample because it is reflected.

Having some background concerning the typical features of time-domain spectra, we now embark on a more formal analysis.

10.4.5 Extracting optical constants – details

Let us now put the discussion on a more quantitative footing. Let us assume the electric field is written as

$$E = E_0 \exp[-i(kz - \omega t)] = E_0 \exp[-i(\tilde{k}(\omega)z - \omega t)]. \tag{10.39}$$

10.4 Time-domain spectroscopy

In the second expression the angular wavenumber is shown explicitly to be a complex number and a function of ω.

We will envisage the sample to be of thickness Δz. We will set the origin of the z axis to be at the front of the sample and so the back of the sample will be at $z = \Delta z$. The electric field at the front of the sample is

$$E(z = 0) = E_0 \exp[-i(-\omega t)], \qquad (10.40)$$

and at the back of the sample

$$E(z = \Delta z) = E_0 \exp[-i(\tilde{k}(\omega)\Delta z - \omega t)] = E(z = 0) \exp[-i(\tilde{k}(\omega)\Delta z)]. \qquad (10.41)$$

Thus the final term, $\exp[-i(\tilde{k}(\omega)\Delta z)]$, represents the change in the electric field on passing through the sample.

Let us imagine that without the sample in the beam, the wave propagates as in a vacuum. (The space may actually be a vacuum, or may optically approach a vacuum, for example, be nitrogen gas, or purged with dry air, but this won't make much difference.) Recall (Equation 6.77) that the angular wavenumber has the relationship

$$\tilde{k}(\omega) = \frac{\omega}{c}\tilde{n}(\omega). \qquad (10.42)$$

In a vacuum, the value of $\tilde{n}(\omega)$ is one, so

$$\tilde{k}_{\text{ref}}(\omega) = \frac{\omega}{c}. \qquad (10.43)$$

We can write the angular wavenumber in the sample in various equivalent forms:

$$\begin{aligned}
\tilde{k}_{\text{sam}}(\omega) &= \frac{\omega}{c}\tilde{n}(\omega) \\
&= \frac{\omega}{c}[n_1(\omega) + in_2(\omega)] \\
&= \frac{\omega}{c}[n_1(\omega)] + i\frac{\alpha}{2} \\
&= \frac{\omega}{c} + \frac{\omega}{c}[n_1(\omega) - 1] + i\frac{\alpha}{2}.
\end{aligned} \qquad (10.44)$$

In the second last step, the definition of α was used, Equation (6.97). In the last step, the term corresponding to the reference phase was separated out. The remaining terms correspond to how the sample differs from free space.

Assuming the field at the front of the sample position is as given in Equation (10.40), the electric field at the back of the sample will be

$$E(\Delta z)_{\text{sam}} = E(0) \exp[-i(\tilde{k}_{\text{sam}}(\omega)\Delta z)] \qquad (10.45)$$

and, without the sample, at the position corresponding to the back of the sample

$$E(\Delta z)_{\text{ref}} = E(0) \exp[-i(\tilde{k}_{\text{ref}}(\omega)\Delta z)]; \qquad (10.46)$$

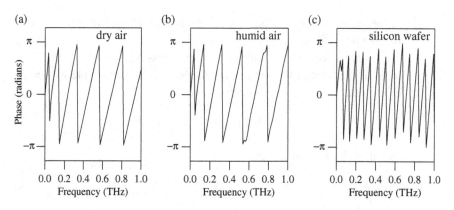

Figure 10.8 Fourier transforms of the time-domain spectroscopy of (a) dry air, (b) humid air and (c) a silicon wafer. The Fourier transform is a complex quantity; here the phase of the complex quantities is shown.

the ratio of these is

$$\frac{E(\Delta z)_{sam}}{E(\Delta z)_{ref}} = \frac{E(0)\exp[-i(\tilde{k}_{sam}(\omega)\Delta z)]}{E(0)\exp[-i(\tilde{k}_{ref}(\omega)\Delta z)]}$$
$$= \exp[-i[\tilde{k}_{sam} - \tilde{k}_{ref}]\Delta z]$$
$$= \exp[-i([\frac{\omega}{c} + \frac{\omega}{c}[n_1(\omega) - 1] + i\frac{\alpha}{2}] - [\frac{\omega}{c}])\Delta z]$$
$$= \exp[-i(\frac{\omega}{c}[n_1(\omega) - 1] + i\frac{\alpha}{2})\Delta z]. \qquad (10.47)$$

So, by looking at the real and imaginary parts of this ratio, the real part of the refractive index and the absorption coefficient may be established.

Usually, E_{ref} and E_{sam} are measured in the time domain as real quantities, $E_{ref}(t)$ and $E_{sam}(t)$. The Fourier transform is taken to yield the complex quantities in the angular frequency domain,

$$\frac{E(\Delta z, \omega)_{sam}}{E(\Delta z, \omega)_{ref}} = \exp[-i(\frac{\omega}{c}[n_1(\omega) - 1] + i\frac{\alpha}{2})\Delta z]. \qquad (10.48)$$

I'll illustrate the outcome of the analysis for the three example time-domain spectra introduced previously, dry air, humid air and silicon wafer (Figures 10.6 and 10.7).

The phase spectra (Figure 10.8) all appear to oscillate. In fact, this can be considered an artefact, as the phase is only shown in the range $\pm\pi$ radians, as usually defined. Another way to present the data would be to add each successive phase segment on the last; that is, rather than reverting to $-\pi$ radians once π radians was succeeded, just continue on to greater numbers. For dry air, Figure 10.8a, the phase would then just continue to increase. What does this mean? It means, for a fixed length of air, as we change the frequency of radiation, the phase change between the start and finish will

10.4 Time-domain spectroscopy

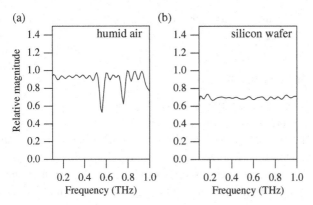

Figure 10.9 Ratios of the magnitudes of Fourier transforms. (a) Ratio of humid air to dry air. (b) Ratio of silicon wafer to dry air.

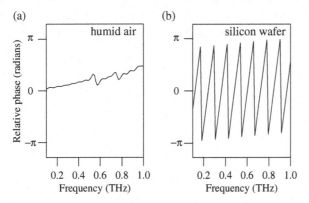

Figure 10.10 Ratios of the phases of Fourier transforms. (a) Ratio of humid air to dry air. (b) Ratio of silicon wafer to dry air.

change systematically. For humid air, Figure 10.8b, the phase changes in the same way, except for some sharp variations, at the characteristic absorption frequencies. For the silicon wafer, Figure 10.8c, the phase change is uniform, as for the dry air, but much more rapid, in view of the higher refractive index.

The results of taking the ratios of the magnitudes are shown in Figure 10.9. The ratio of the humid air to the dry air, Figure 10.7b divided by Figure 10.7a, is given in Figure 10.9a. The magnitude ratio is close to unity, except where characteristic absorptions occur, such as at 0.5 THz. The ratio of the silicon to the dry air, Figure 10.7c divided by Figure 10.7a, is given in Figure 10.9b. There are no characteristic absorptions, but the overall magnitude ratio is about one half, attributable to the refractive index mismatch between silicon and dry air.

The results of taking the ratios of the phases are shown in Figure 10.10. The ratio of the humid air to the dry air, Figure 10.8b divided by Figure 10.8a, is given in Figure 10.10a. The phase shows a characteristic wiggle at the resonances, such as at 0.5 THz. The ratio of the silicon to the dry air, Figure 10.8c divided by Figure 10.8a, is given in Figure 10.10b. There are no characteristic wiggles, but a steady mismatch due to the refractive index difference.

10.4.6 Time-resolved spectroscopy

The concept of splitting the excitation beam into parts that gate the emitter and detector may be taken a step further: the same source beam may be used to excite the sample under study. This technique is known as **time-resolved spectroscopy**, terahertz time-resolved spectroscopy or time-resolved terahertz spectroscopy.

To implement time-resolved spectroscopy, an additional beamsplitter and additional delay stage are required. Typically, the excitation beam is split into two components, one of which is directed onto the sample, say, to promote electrons into the conduction band. The second component is then used as the source of a conventional time-domain spectroscopy system. By varying the delay between the first and second beam, the terahertz spectrum of the sample can be collected at different stages after excitation. This can yield information about carrier dynamics on the sub-picosecond time scale.

10.5 Summary

10.5.1 Key terms

dispersive spectroscopy, 210   interferometry, 216   time-domain spectroscopy, 223

10.5.2 Key equations

grating equation	$mc = fa(\sin\theta_m + \sin\theta_i)$	(10.2)
chromatic resolving power	$R = Nm$	(10.16)
interferogram transform	$B(\sigma) = \int_0^\infty I(x)\cos(2\pi\sigma x)dx$	(10.26)
time-domain spectroscopy	$\dfrac{E(\Delta z, \omega)_{\text{sam}}}{E(\Delta z, \omega)_{\text{ref}}} = \exp[-i(\dfrac{\omega}{c}[n_1(\omega) - 1] + i\dfrac{\alpha}{2})\Delta z]$	(10.48)

10.6 Table of symbols, Chapter 10

General mathematical symbols appear in Appendix B.

Symbol	Meaning	Unit
a	grating period	m
b	ramp length	m
c	lightspeed	m Hz
d	path length	m
d_m	path length for m cycles	m
E_0, E_1, E_2	electric field amplitude; total, first wave, second wave	V m^{-1}
f	frequency	Hz
f_1	frequency of first wave	Hz
f_2	frequency of second wave	Hz
I_0, I_1, I_2	electric field intensity; total, first wave, second wave	J m^{-2} Hz
$\tilde{k}$	complex angular wavenumber	m^{-1}
$\tilde{k}_{\text{ref}}$	complex angular wavenumber, reference	m^{-1}
$\tilde{k}_{\text{sam}}$	complex angular wavenumber, sample	m^{-1}
n	refractive index	[unitless]
$\tilde{n}$	complex refractive index	[unitless]
n_1	real part of refractive index	[unitless]
n_2	imaginary part of refractive index	[unitless]
N	total number of diffracting elements	[unitless]
R	chromatic resolving power	[unitless]
t	time	Hz^{-1}
t_s	time in sample	Hz^{-1}
t_v	time in vacuum	Hz^{-1}
x	path length	m
z	path length in sample	m
α	blaze angle	rad
α	absorption coefficient	m^{-1}
δ	phase difference	rad
θ	exit angle	rad
θ_i	incidence angle	rad
θ_m	exit angle corresponding to order m	rad
λ	wavelength	m
λ_1, λ_2	wavelength; first, second wave	m
σ	wavenumber	m^{-1}
σ_1, σ_2	wavenumber; first wave, second wave	m^{-1}
ω	angular frequency	rad Hz

11 IMAGING

This chapter calls on maths, but the maths is relatively elementary. The first four sections only require simple geometry. The final sections refer to Fourier transforms, but only at a descriptive level.

Sight is the most complete of our senses. Our eyes detect light. So our eyes are photon detectors. There's more: our eyes distinguish light of different colours. So our eyes are spectrometers. There's more: our eyes tell the direction the light is coming from. So our eyes are imaging devices. There's more: between them, our eyes let us build up a three-dimensional image of the scene we are viewing.

We extend our vision using instruments. For example, the telescope lets us see the distant; the microscope, the small.

We can extend our vision to other parts of the electromagnetic spectrum. To do this, we need an instrument sensitive to invisible radiation that converts it to something we can see. X-rays are an example. An x-ray viewer records x-rays arriving from different places and presents this in a way the eye can see. At first, photographic film was used to display x-rays. Now a computer monitor is standard. In principle, the x-rays could be separated according to frequency, but this is not usually done in practice. By taking multiple x-ray images from different angles, a three-dimensional x-ray image may be built up. This process is called computer-axial tomography or computer-aided tomography (CAT) or simply computer tomography (CT). You may have seen, or even been in, a tunnel-like machine called a CAT scanner.

In the same way, we can now see with terahertz-frequency radiation. Using terahertz sources and detectors, we can build up the pattern of the radiation coming from different places and so construct an image. We can do this using different frequencies, the equivalent of taking photographs of different hues. We can do this from many angles, and reconstruct a three-dimensional model. This chapter sets out the principles behind these techniques.

An *image* is a two-dimensional pattern of light. The simplest images offer only two options at each point: black or white. Black and white can be represented by two numbers, such as 1 and 0. Using this representation it is easy to store the image on a computer and it is easy to carry out mathematical operations on it. Next in sophistication comes the grey-scale image. Here the extremes of black and white are bridged by a series of shades of grey. On a computer, the number of shades of grey is usually chosen to be a power of two: 2 (black and white), 4, 8, 16, 32 or 64. The eye can not distinguish more than about two dozen shades of grey, so there is little need to extend this further. Next in sophistication comes the colour image. To give complete colour information, a full spectrum must be specified at each image point. In practice, a good approximation to human vision is made by mixing in the required amounts of only three colours, red, green and blue (RGB). As with shades of grey, the human eye can only distinguish a small range of colours. It is quite practical to display on a computer monitor many more than the number of distinct colours perceptible.

So far, I have been discussing a single point on the image. In principle an image may have an infinite number of points. In practice this is pointless as the limits of the human eye can easily be accommodated by a finite number of points. For example, a PAL television has 625 lines. The separate lines may be discerned if you are close to the screen, but at normal viewing distance they merge to an apparent continuum.

Let us now introduce the dimension of time. Although in principle successive images could be at any time separation, there is no need to do so if a human observer is the end-user. Because of the persistence of vision, successive images at small intervals cannot be distinguished by you or by me. Television and movies exploit our inability to distinguish images close in time and are typically made with 24 or 25 or 30 frames per second (fps).

To sum up, although we can conceive of an image having a large or even infinite number of points, comprising a large amount of spectral information, and being changed continuously in time, human sight has limits. Thus, in practice, we can record the image with only finite spatial, spectral and temporal resolution.

Today, most images are *digital*. The location of a point on the image, the colour at that point, the timing relative to other images are all represented by numbers. Most images today are recorded, stored and transmitted in digital format.

Imaging methods in the terahertz region share similarities with imaging methods in other spectral regions, such as conventional photography using visible light. Some photographs are of things that emit light, such as the sun or the stars or fireworks. This is referred to as *passive imaging*. Passive imaging is also available in the terahertz regime. Conventional photography usually involves light being reflected or scattered from a scene. In outdoor photography, the source of light is usually the sun. In studio photography, lamps are used. Illuminating an object with terahertz radiation and monitoring the reflected radiation is *reflection imaging*. Likewise, monitoring the radiation passing through an object is *transmission imaging*. This is the usual mode of x-ray imaging, although it is seldom employed in conventional photography.

The angular resolution is normally limited by the aperture of the imaging device (the size of a camera lens, for example) and the frequency of the radiation used. By placing a sensor extremely close to an object, this limitation can be overcome. This method is called *near-field imaging*. By combining a set of transmission images taken from different angles it is possible to reconstruct the internal structure of an object. This method is known as *tomography*.

Images in your eye or in a camera are acquired as a whole. The light is received simultaneously over an array. Different elements of the array receive light from different directions. This is called a *focal plane array*. Images on a television screen or computer monitor are built up line by line or point by point. To sweep across one line, then sweep across a line below, then across the line below, is to *raster*. Often in the terahertz regime, images are built up this way, line by line, point by point.

Learning goals

By the time you finish this chapter you should be able to give an account of terahertz

- passive imaging,
- raster imaging (both transmission and reflection),
- focal plane array imaging,
- near-field imaging,
- tomography.

11.1 Passive imaging

We speak of **passive imaging** when we simply collect radiation. We do not supply radiation to the object under scrutiny, merely detect the radiation coming from it already. Usually, we think of the object being an emitter of radiation, although it may merely be reflecting or scattering radiation from another source outside our control. The point is, we do not supply terahertz radiation, only detect it.

Passive imaging is often, although not essentially, connected with us not manipulating the object under study. It is like a news camera operator recording action over which she has no control, as opposed to a studio photographer who asks her subject to look left and to smile.

Passive imaging is used for imaging distant objects, such as astronomical sources, or for surveillance, where the observer does not want to give away his presence by shining radiation on the target.

In observing everyday objects as terahertz emitters, the main contrast mechanism is temperature difference. We have seen that objects at any temperature above absolute zero will emit radiation of all frequencies, including terahertz frequencies. For example, a person at normal body temperature will radiate more energy per surface area than the walls of a house at room temperature, and so an image of the person in front of the cooler wall can be obtained. If the person happened to be outside on a rather hot day and the wall was at body temperature, there would be no contrast, and the person would be 'invisible'. On the other hand, there is good contrast against the sky, which has a much lower effective temperature than a person.

In the discussion of contrast based on temperature differences, I have neglected emissivity. Two objects at the same temperature will radiate at different rates if their emissivities are different. It is quite feasible to distinguish objects at the same temperature but with different emissivity using passive terahertz imaging.

The essential optics of a passive imaging system appear in Figure 11.1. The rays coming from the object reach the terahertz detector through an aperture of opening size D. This diameter is variously known as the aperture stop, the aperture diameter or the

11.1 Passive imaging

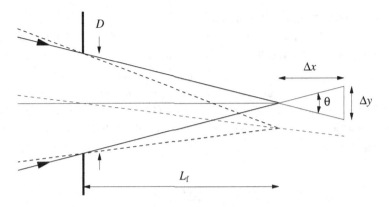

Figure 11.1 Passive imaging. The aperture stop, of length D, is usually defined by a focussing optical element, such as a mirror or lens. The focal length, L_f, is the distance from the aperture to the focus. The f-number $N \equiv L_f/D$. (The figure is drawn for $N = 2$.) The angular field of view, or full cone angle, θ, is approximately $D/L_f = 1/N$.

limiting aperture. It is typically the diameter of the focussing lens or mirror collecting the rays. Of course, the aperture does not have to be circular in shape, but for simplicity we will assume it is.

A second characteristic length in the imaging system is the distance from the aperture to the detector, the focal length, L_f. (In optics, the focal length is often denoted f, but in this book we reserve f to denote that most significant of quantities, frequency.)

The ratio of the two characteristic distances just defined is the **f-number**, denoted N. (It is also denoted $f/\#$.) The f-number is simply the ratio of the focal length to the aperture diameter,

$$N \equiv \frac{L_f}{D}. \tag{11.1}$$

Example 11.1 Camera lens

A typical camera has a focal length of 50 mm and an f-number of 2. The lens aperture is then 25 mm in diameter. For a given focal length, a larger f-number means a smaller aperture. In a camera, the focal length is typically more or less fixed and the f-number is increased by reducing the lens aperture using a variable iris.

Exercise 11.1 f-number of mirror

An off-axis paraboloidal mirror used for focussing a terahertz beam has a radius of 25.4 mm and an effective focal length of 101.3 mm. What is the f-number of this mirror? ∎

Exercise 11.2 f-number of lenses
A series of TPX lenses has focal lengths of 25, 50, 100 and 200 mm. Each lens is of diameter 25.4 mm. What is the $f/\#$ of each lens? ∎

From this definition of N, we can write an expression for the full cone angle of the rays entering the aperture and headed for the focus. The half cone angle has the tangent of $D/2$ on L_f, so the full cone angle is $2\arctan(D/2L_f)$. In the case of a small angle, we can replace the tangent of the angle with the angle itself. In this approximation the full cone angle is

$$\theta \sim \frac{D}{L_f} = \frac{1}{N}. \tag{11.2}$$

Another quantity used in this context is the *numerical aperture*. This is defined by

$$NA \equiv n \sin \frac{\theta}{2}. \tag{11.3}$$

Here n is the refractive index of the uniform material the optical system inhabits. When the angle is small, and so approximately equal to the sine of itself, and if as well the refractive index is one, then from Equations (11.2) and (11.3),

$$NA \sim \frac{1}{2N}. \tag{11.4}$$

A key limitation to terahertz imaging is its resolution. We have discussed before spectral resolution, meaning the smallest frequency difference that can be detected. Now we are not talking about distinguishing frequencies, but distinguishing different positions. We take two point sources and ask the question, how close can they be for us to distinguish them in a terahertz image? The **angular resolution** is the difference in angle between two point sources that are just resolved. Angular resolution is convenient to use when discussing distant objects, such as astronomical sources. The **lateral resolution** is the distance between two point sources that are just resolved. Lateral resolution is convenient to use when discussing samples in the laboratory. The angular resolution and the lateral resolution can be deduced from each other if the distance to the object is given.

One criterion used to specify when two point objects are just resolved is the Rayleigh criterion. The Rayleigh criterion states two point objects are just resolved when the first zero in the diffraction pattern of one object falls on the maximum in the diffraction pattern of the second object. (There are alternative criteria, such as the Sparrow criterion.) The **Rayleigh angle** gives the angular separation of two point objects according to the Rayleigh criterion; in terms of wavelength

$$\theta_{\text{Rayleigh}} \sim \frac{1.22\lambda}{D}; \tag{11.5}$$

in terms of frequency,

$$\theta_{\text{Rayleigh}} \sim \frac{1.22c}{fD}. \tag{11.6}$$

The expressions are obtained from the theory of diffraction, but the derivation is not given here. To improve the resolution, we can either increase f or increase D. (Can't do much with c.)

Numerically, if the aperture diameter is given in millimetres and the frequency is given in terahertz, the Rayleigh angle, in degrees, is

$$\theta_{\text{Rayleigh}} \text{ [in degrees]} \sim \frac{20.96}{f \text{ [in terahertz]} D \text{ [in millimetres]}}. \tag{11.7}$$

Example 11.2   Rayleigh angle

Using an aperture of $D = 10$ mm and radiation of $f = 1$ THz, the Rayleigh angle is about 2.096 degrees. So objects at an angular separation of 3 degrees are distinguished, but not objects separated by only 1 degree. At 0.1 THz, the Rayleigh angle is about 21 degrees, at 10 THz about 0.21 degrees, for the same opening.

Exercise 11.3 Rayleigh angle

Calculate the Rayleigh angle for optics with 25.4 mm diameter at the frequencies of 0.1, 1.0 and 10 THz. ∎

Exercise 11.4 Rayleigh angle

What is the Rayleigh angle for the frequencies of 0.1, 1.0 and 10 THz for optics of 1 mm diameter? ∎

I have given an expression for angular resolution; now let's turn to lateral resolution. For an object at a distance L_o, the Rayleigh angle θ_{Rayleigh} corresponds to a lateral distance $\theta_{\text{Rayleigh}} \times L_o$. As the object is moved closer, this lateral distance decreases accordingly. The closest we can bring the object is right up to the imaging aperture, when $L_o = L_f$. The lateral resolution is then as small as it gets. So the minimum lateral separation is

$$D_{\text{Rayleigh}} \sim \frac{1.22c}{fD} \times L_f = \frac{1.22c}{f} N. \tag{11.8}$$

Hence the lateral resolution may be improved by reducing the f-number, but it is difficult to engineer an f-number below about 1. Another way to improve the lateral resolution is to increase the frequency f.

Numerically, the minimum lateral resolution is

$$D_{\text{Rayleigh}} \text{ [in millimetres]} \sim \frac{0.366N}{f \text{ [in terahertz]}}. \tag{11.9}$$

Example 11.3 Lateral resolution
For an f-number of 1, and a frequency of 1 THz, the lateral resolution is about 0.4 mm. It is difficult to arrange an optical system with an f-number lower than 1. So the practical limit for the resolution is about the size of the wavelength; in this case, about 0.4 mm for a wavelength of about 0.3 mm.

Exercise 11.5 Lateral resolution
For a frequency of 0.1 THz, what is the lateral resolution, assuming an f-number of 4? ■

Exercise 11.6 Minimum lateral resolution
Given an f-number of 1, and a frequency of 10 THz, what is the minimum lateral resolution? ■

A consequence of diffraction is that the image of a point source is not a point at the focal plane but rather smeared over a circle of radius $\theta_{\text{Rayleigh}} L_f$. Now employing geometric optics, if we move the focal plane a distance Δx from the focus, the spot size will be $\Delta y = \Delta x D/L_f$. Moving the focal plane forward or back by an amount such that $\Delta x D/L_f = \theta_{\text{Rayleigh}} L_f$ gives the **depth of focus** as

$$\Delta_{\text{focus}} \sim \frac{2.44c}{f} N^2. \tag{11.10}$$

The depth of focus is the **axial resolution** at the focal point: moving the detector this far will not appreciably change the image size.

We can translate this result to the object plane. That is, we can calculate how far the object can move towards or away from the detector without the image changing significantly. This quantity is the axial resolution at the object plane; more often it is called the **depth of field**. We can calculate the depth of field directly from the depth of focus by dividing by the axial magnification of $(L_f/L_o)^2$. (This expression for the axial, or longitudinal, magnification is the square of the expression for the transverse magnification; the expressions can be derived from geometrical optics but the derivation is not given here.) The quantities L_f cancel out, leaving

$$\Delta_{\text{field}} \sim \frac{2.44c}{f} \left(\frac{L_o}{D}\right)^2. \tag{11.11}$$

Numerically,

$$\Delta_{\text{field}} \text{ [in millimetres]} \sim \frac{0.731}{f \text{ [in terahertz]}} \left(\frac{L_o}{D}\right)^2, \tag{11.12}$$

where L_o and D are measured in the same units.

Example 11.4 Depth of field

An object at 10 cm distance from a 1 cm aperture is imaged at 1 THz. The depth of field is about 73 mm. This quantity is important when imaging thick objects, and so important in tomography.

Exercise 11.7 Depth of focus, depth of field

A sample is placed at the focus of a mirror with 101 mm focal length. The entrance aperture to the detector is 12.7 mm in diameter. For imaging at 3 THz, (a) what is the depth of focus? (b) what is the depth of field? ∎

Let's suppose that a terahertz detector is at the focus in Figure 11.1 and had sufficient time to acquire data. We now have one point, or pixel, of the image. How do we get more image data? For example, how do we obtain data from the object located in the direction of the dashed lines in Figure 11.1?

There are in essence two different ways to build up an image. One is to change the pointing of the whole optical system. So, in Figure 11.1 we would rotate the entire optical system clockwise until it pointed in the direction of the dashed line. In practice, it is usually not the whole optical system, but rather a beam-steering mirror or lens that rotates, while other parts, such as the detector, remain in place. The second approach to build up the image is to set a second detector in the position of the focus of the dashed lines in Figure 11.1. Better still, place a whole lot of detectors on the focal plane. This creates a focal plane array, the topic of Section 11.3. Arrays of eight elements are common, and larger ones are being developed. A further possibility is to combine the two approaches, and scan the optics while using a focal plane array.

Passive imaging systems have been demonstrated around 0.1 THz. They are used at lower frequencies still (around 0.035 THz; in the realm of microwaves) in aircraft navigation, since microwaves are not absorbed as strongly by fog and mist as is visible light. Passive imaging finds applications in security systems. For example, a person will emit terahertz radiation, even through clothing; a concealed weapon may be apparent as a relatively colder part of the image.

11.2 Raster imaging

11.2.1 Transmission

A common way to collect terahertz images in the laboratory is shown in Figure 11.2. The sample is examined in transmission. A data point is taken, then the sample moved a small distance in the x direction. Another data point is taken. Then the sample is moved again. In this way a line of data is collected. The sample is then moved a small distance in the y direction. Then the process of stepping along in the x direction is repeated. The word **raster** describes this motion, building up the data point by point, line by line.

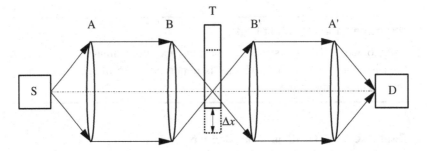

Figure 11.2 Transmission imaging. S is the source, D is the detector. A is a collimating optical element and A' a condensing one; B is a condensing element and B' a collimating one; these are typically mirrors or lenses. The sample, examined in transmission, is marked T. It is shown translated in the x direction. It can also be translated in the y direction. In this way, piece by piece, an image is built up.

In Figure 11.2 the sample is the moving part. Another approach is to leave the sample stationary and scan the beam across it. The collection of the transmitted beam has to be accomplished synchronously. This is generally more difficult to implement than moving the sample, but is useful if the sample is large or heavy.

Many different sources are used in transmission raster imaging, including pulsed terahertz radiation from photoconductive, optical rectification and electro-optical sources and continuous terahertz radiation from Gunn diodes, far-infrared lasers, quantum-cascade lasers, terahertz parametric oscillators and backward-wave oscillators. Likewise, many different detectors are used. To detect continuous radiation, bolometers, pyrometers and Schottky diodes are common. To detect pulsed radiation coherently one of the time-domain spectroscopy detectors, based on photoconductive or electro-optical detection, is used.

The signal collected from a continuous source is usually simply the terahertz power, although it is possible to extract the phase from gated continuous methods, for example, by using photoconductive emitters and detectors.

Much richer information is available using time-domain methods. The whole spectrum is available at each pixel. So it is possible to generate an image at every separate frequency in the frequency-domain spectrum. The time-domain method yields not just one image, but an image at each frequency. For example, if the sample is known to have a resonance at a particular frequency, an image can be retrieved for that frequency.

Rather than use all the time-domain data (or all the transformed data in the frequency domain), simpler approaches have been taken. For example, images may be constructed using the peak of the time-domain signal, the peak-to-peak (that is, maximum to minimum) of the time-domain signal, the time delay of the peak of the time-domain signal (proportional to the phase delay, provided the refractive index is uniform) or the peak of the frequency-domain signal.

Transmission images have been made of chocolate bars, leaves, credit cards, biological sections, dollar bills, fluorescent markers and teapots. Demonstrations have been made of detecting foreign or concealed objects, such as needles in milk powder, metal or ceramic blades hidden under cloth or plastic and powders of various sorts concealed in envelopes.

Exercise 11.8 Raster time

An image of size 1024 by 768 pixels is being accumulated by raster scanning. A lock-in amplifier with time constant 30 ms is being used and the dwell time at each point is 5 time constants. How long will it take to collect the image? ∎

11.2.2 Reflection

An arrangement similar to Figure 11.2 may be used for reflection imaging and the same principles apply. As with transmission raster imaging, diverse source and detectors have been used for reflection raster imaging. Commonly, the reflected amplitude is measured.

Using time-domain spectroscopy, a particular modality named **time-of-flight imaging** has been developed using reflection imaging. The principle is illustrated in Figure 11.3. The reference ray, (a), reflects off a metal surface which we can take as the top

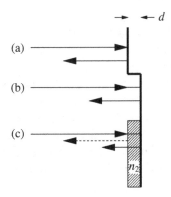

Figure 11.3 Reflection imaging. The thick line represents a highly reflective surface, such as a metal. The shaded region represents a dielectric, characterised by refractive index n_2. (The diagram is drawn for $n_2 = 1.5$.) Ray (a) reflects off the first metal surface; take this as the reference ray. Ray (b) reflects off a recessed metal surface; this ray is delayed with respect to ray (a). Ray (c) reflects from the recessed metal surface, and also passes through the dielectric twice. It is delayed even more than ray (b). Because of the dielectric mismatch with air, there will also be reflection from the top of the dielectric (dashed line), but this will be weaker than the reflection off the metal.

of the sample. A second ray, (b), reflects off a recessed section. It will travel farther, both in its outward journey and on the way back. If the recess is of depth d, the additional distance travelled is $2d$, and the additional time taken is $2d/c$. The time delay can be measured directly in the time-domain spectrum. If the sample is entirely made of the same metal, the delays as measured in the time-domain spectra can be directly and quantitatively used to map the depth profile.

Perhaps a ray passes through a dielectric coating on the metal, as does ray (c) in Figure 11.3. Two return pulses will be evident in the time-domain spectrum. First, at the same time delay as for ray (a), there will be a weak reflection from the dielectric/air interface. Second, there will be a pulse from the metal substrate. Because of the presence of the dielectric, this will be delayed even further than ray (b). The additional distance travelled is the same, $2d$, but the optical path is larger by the factor of the refractive index and is $2dn_2$. This gives a delay of $2dn_2/c$.

Reflection imaging is usually more difficult in practice than transmission imaging. It comes into its own when the sample is not very transmissive: if it is very reflective, as in the case of a metal, or very absorptive, as in the case of materials containing a lot of water, such as biological samples. For example, terahertz radiation does not penetrate far in human tissue, typically less than 1 mm, so can not be used to image internal structures, as x-rays can. In medical applications, terahertz radiation is largely used in reflection, and is sensitive to the top layer of skin. As examples, reflection imaging has been used to investigate human skin cancers and burns to chicken and pig flesh.

Exercise 11.9 Time of flight

A recess 5 mm deep in a metal is being probed by reflection spectroscopy. (a) What is the time delay between the beam from the recess and one from the top surface? (b) What is the time delay if the recess is filled with epoxy of refractive index 1.5? ∎

11.3 Focal plane array imaging

The concept of collecting information from many parts of an object simultaneously by a **focal plane array** was introduced in Section 11.1. Rather than scanning the sample or the beam, as in Sections 11.2.1 and 11.2.2, separate pieces of information are collected by an array of detectors.

The array of detectors may consist of bolometer elements constructed using superconductors or electronic detectors comprising high-electron mobility transistors. At present, it is difficult to fabricate either in large arrays.

An ingenious way to realise a focal plane array for terahertz detection is shown in Figure 11.4. Rather than have an array of terahertz detectors, this method uses an electro-optic crystal to convert the terahertz signal into the rotation of the polarisation of a near-infrared probe beam. The probe beam is imaged using a CCD camera readily available at near-infrared wavelengths. This method requires a strong source of terahertz radiation, as all the pixels are illuminated simultaneously. Nevertheless, video rate movies have been achieved using such a system.

11.4 Near-field imaging

It is difficult to image an object smaller than the wavelength of the light being used. This follows from Equation (11.8). For example, using visible light, of wavelength about half a micrometre, it is difficult to resolve objects smaller than about half a micrometre across. With light of 1 THz, it is difficult to resolve objects smaller than about 300 μm in size.

Here is a simple analogy. Insect screens are designed to keep out insects, and the minimum screen size depends on the size of the bug. For example, a mosquito net has

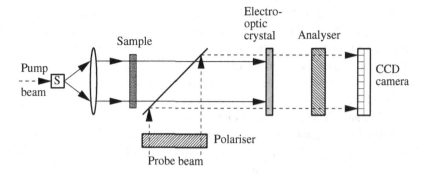

Figure 11.4 Focal plane array imaging. Synchronised near-infrared pulses (dashed lines) are used to excite the terahertz source, S, and to detect the terahertz radiation in an electro-optic crystal. The terahertz beam (full lines) illuminates the whole of the sample at once. After passing through the sample, this beam illuminates a large area of the detector crystal. The probe beam, shown slightly larger than the terahertz beam, illuminates the same area of the electro-optic crystal. The probe beam continues to the analyser (the terahertz beam is no longer needed) and the rotation of polarisation induced by the terahertz beam in the electro-optic crystal becomes evident. The probe beam is detected over a large area by the CCD camera located in the focal plane.

a smaller mesh than a fly screen, since mosquitoes are smaller than flies. (I heard of a man with two types of fly screens on his windows, one with a large mesh to keep out large flies and one with a small mesh to keep out small flies.) When I first learned of a tick fence, I envisaged a rather low wall with very fine mesh to prevent the passage of those minute creatures; it turns out the fence is a rather larger construction, to stop the passage of the cattle carrying the ticks. But you get the idea: although a security screen, with bars 10 cm apart, is impervious to a person, it is easily breached by an ant.

Consider a meshing with wires about every 5 cm, such as is found around a tennis court to prevent tennis balls escaping. A tennis ball hit at the net will bounce off. For tennis balls, the mesh is a perfect reflector. There is no transmission of tennis balls through the mesh. The chance of sending a larger object – say, a beach ball – through the mesh is even more remote. To find out about the structure of the mesh we want something of a smaller dimension, say, a pea. Shooting peas at the mesh we would find not all are reflected. Some are transmitted. Investigating this on a quantitative basis allows us to measure the size of the mesh.

Let's consider the mesh around the tennis court and the tennis ball more closely. And I do mean closely. If we held a tennis ball very close to the mesh we would find in some places, where the ball was resting right in the centre of a gap, it would protrude slightly through the mesh. At another place, where the leading edge of the ball rested on a wire rather than the gap, it would not protrude beyond the mesh at all. So the transmission is not quite zero, if we look closely enough; but the distance the ball gets through the net is rather less than its diameter.

Throwing a tennis ball at a mesh or a wall and seeing where it ends up, whether it passes through or bounces back, is like 'far-field' imaging. Pressing a tennis ball against a mesh and seeing if any at all protrudes is like 'near-field' imaging.

More precisely, **far-field imaging** refers to detecting radiation at a distance much greater than the wavelength. All the imaging methods we have considered thus far are far-field. Detecting radiation at a distance of the order of the wavelength or less is **near-field imaging**. This is what we turn to now.

Near-field imaging at 0.1 THz means nearer than about 3 mm; at 1 THz, nearer than about 300 μm; at 10 THz, nearer than about 30 μm. Far-field imaging applies to distances much greater than these.

Let's put the discussion on a more quantitative footing. Consider an electromagnetic wave whose spatial part is represented by the equation $E = E_0 \exp[i(k_x x + k_y y + k_z z)]$ encountering an aperture of diameter a in the x-y plane. The components in the x and y directions must be greater than $2\pi/a$, that is, $k_x > 2\pi/a$, $k_y > 2\pi/a$. The total magnitude of the vector in terms of its components is

$$k^2 = k_x^2 + k_y^2 + k_z^2. \tag{11.13}$$

Rearranging to make k_z the subject,

$$k_z^2 = k^2 - k_x^2 - k_y^2. \tag{11.14}$$

Now writing $k = 2\pi/\lambda$ and imposing the condition for the minimum values of k_x and k_y,

$$k_z^2 = \left(\frac{2\pi}{\lambda}\right)^2 - \left(\frac{2\pi}{a}\right)^2 - \left(\frac{2\pi}{a}\right)^2 = \left(\frac{2\pi}{\lambda}\right)^2 - 2\left(\frac{2\pi}{a}\right)^2. \tag{11.15}$$

This may be written as

$$k_z^2 = -(2\pi)^2 \left[2\left(\frac{1}{a}\right)^2 - \left(\frac{1}{\lambda}\right)^2\right]. \tag{11.16}$$

The equation has been written in this way to ensure the term in the square brackets is positive for the case we are considering, $a < \lambda$. This has the effect of bringing a negative sign to the front of the equation. The square of k_z is a negative number. Hence k_z must be imaginary. Taking the square root of both sides of the equation and introducing $i = \sqrt{-1}$,

$$k_z = 2\pi i \left[2\left(\frac{1}{a}\right)^2 - \left(\frac{1}{\lambda}\right)^2\right]^{1/2}. \tag{11.17}$$

If $a \ll \lambda$, then $(1/a)^2 \gg (1/\lambda)^2$, and we can neglect the term involving λ to obtain

$$k_z = \frac{\sqrt{8}\pi}{a} i. \tag{11.18}$$

Substituting this value for k_z into the equation of the wave we have, for the z dependence,

$$E(z) = E_0 \exp[ik_z z] = E_0 \exp\left[i\left(\frac{\sqrt{8}\pi}{a} i\right) z\right] = E_0 \exp\left[-\frac{\sqrt{8}\pi}{a} z\right]. \tag{11.19}$$

11.4 Near-field imaging

This does not represent a propagating wave, but rather a decaying or evanescent wave. It penetrates the barrier, but fades away in doing so with characteristic decay length

$$\delta = \frac{a}{\sqrt{8\pi}}. \qquad (11.20)$$

Hence, we can image the aperture, provided we detect the radiation very close to the aperture itself. By very close, I mean no more than about the distance δ, which is of the order of the aperture width a, from the aperture. If you think back to the tennis ball being pushed against the mesh, only a small part of the ball protrudes through the mesh.

There are many experimental configurations for near-field imaging. Three are shown in Figure 11.5.

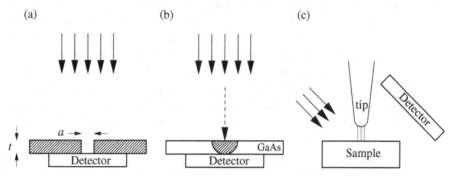

Figure 11.5 Near-field imaging. (a) Using a fixed aperture. (b) Using a dynamic aperture. (c) Using a tip.

Near-field imaging using a fixed aperture, Figure 11.5a, places the detector behind a sub-wavelength aperture, typically a gap in a metal mask. The nearer the detector to the aperture the better, and, as we see from Equation (11.19), the detector should be no farther away than about the length of the aperture itself. Here we run into a problem. We want to bring the detector as close as we can to the aperture opening, but to do so we need to decrease the thickness, t, of the metal mask. We have seen that terahertz radiation penetrates even metals; the penetration depth is of the order of 100 nm. So if the thickness of the metal mask is of this order, significant amounts of terahertz radiation will get through the mask to the detector, nullifying its role as an aperture. This sets a practical limit for this sort of physical aperturing in the range 1 to 10 μm.

Rather than a permanent, physical aperture, a dynamic, or transient aperture is realised in the arrangement in Figure 11.5b. For example, the detector can be covered with a layer of photoconductive material, such as GaAs. Usually, GaAs will transmit terahertz radiation. However, if it is illuminated with above-bandgap radiation, such as from a near-infrared laser, electron-hole pairs are produced at the laser spot. This produces a highly conductive or metallic region at the laser spot. This region will reflect terahertz radiation, and so prevent its transmission. Thus is made an anti-aperture. The spot size is of the order of 10 µm in diameter, and so the resolution achieved is of this order. An image is built up as the laser beam scans across the photoconductor surface, sweeping out a transient, localised metallic spot.

A third practical method of near-field imaging is shown in Figure 11.5c. It is related to the methods used in the field of scanning probe microscopy, which embraces scanning tunnelling microscopy and atomic-force microscopy. The idea is that a tip scans across the sample surface and an image is built up by recording some interaction of the tip and the sample. Depending on the interaction, the resolution can be very high, of the order of 10 nm or so. If the interaction depends on electron tunnelling, for example, it is very sensitive to the distance between the tip and the surface, exponentially, so most of the interaction occurs at the very end of the tip. A resolution can be obtained far better than the physical size of the tip. The tip-surface interaction is then used to modulate the response to terahertz radiation. The method is typically used in reflection, and resolutions of the order of 10 nm have been achieved.

Further methods to achieve sub-wavelength resolution are available for specific materials. For example, we have seen that in generating terahertz radiation by the photoconductive, electro-optical or optical rectification methods, an excitation spot of transverse dimension of 10 µm or so is used to produce radiation of much greater wavelength, say 300 µm. Relative to the wavelength, we can regard these as point sources. The excitation point can be scanned across a material that emits terahertz radiation through any of these processes, and differences in the terahertz radiation emitted allow an image to be built up. Likewise, the distance into such an emitter that the excitation radiation penetrates, typically 1 µm, is very much less than the wavelength of the terahertz radiation emitted. With suitable emitters as samples and appropriate geometries, depth resolutions much smaller than the emission wavelength are obtained.

11.5 Tomography

In previous chapters, we have considered the terahertz signal from a single spot on a sample. In this chapter, we have considered the terahertz signal at different locations. By scanning the terahertz beam along a line on the sample, we can build up a line profile. Combining a number of line profiles, we can build up a two-dimensional image. We can go further, and combine two-dimensional images to build up a three-dimensional image.

Three-dimensional imaging is part of normal human vision. Two eyes – binocular vision – leads to a perception of depth. A hologram is one way to use an essentially two-dimensional image to produce a three-dimensional view. '3D' movies and television rely on sending slightly different two-dimensional images to each eye, resulting in the perception of a three-dimensional scene.

The three-dimensional images generated by 3D television or movies tend to be of the external form of the object. Terahertz three-dimensional images can be made of the external form of an object, for example, by using the time-of-flight method in reflection. Three-dimensional images may also be made of the internal structure of an object. For example, x-rays are used to image the internal structure of the human body.

11.5 Tomography

'Tomo-' means slice and tomography means an image of a slice, or a plane section. In x-ray tomography, slices or sections through the human body are generated. These may be stacked together to give a three-dimensional representation of the internal structure. In a body scanning x-ray machine, it is usual for the scannee to lie on a bed while the x-ray sources and detectors move on a ring circling the body. The bed then moves through the source/detector hoop and the data are built up slice by slice, plane by plane. Since the data are then processed by computer to reveal the internal structure, the method is called **computer tomography** or **computer-aided tomography**. Since the sections are perpendicular to the long axis of the human body, the method is also called **computer-axial tomography**.

The modern method of tomography is based on Fourier transforms, as illustrated in Figure 11.6. First, parallel beams of light are sent through the object from a particular direction, in Figure 11.6a perpendicular to the x axis. The radiation is detected on the opposite side. The data are called the *projection* and written as $p(x)$. The next step is to plot these data on the projection plane, Figure 11.6b. For the square object shown, the function on the projection plane would be zero for $x < 0$, then constant between $x = 0$ and $x = w$, where w is the width of the square, then zero for $x > w$. This is not represented in detail in Figure 11.6b, but a dot-dash line shows where the data lie, the data themselves coming out of the plane in the z direction, out of the page. Next, a one-dimensional Fourier transform of the data is calculated. The result is plotted in the Fourier plane and lies along the k_x axis. This is shown in Figure 11.6c. The whole process is repeated for other angles of projection. One of these is shown in Figures 11.6d, 11.6e, 11.6f. After doing this for many angles of projections, a dense set of data points is built up in the Fourier plane.

Now, here's the rub. Taking the inverse two-dimensional Fourier transform of the data in the Fourier plane leads back to the original object in real space. Bingo.

Perhaps you missed it, wondering how the rabbit sprang from the hat. In going from (b) to (c), or from (e) to (f), a one-dimensional Fourier transform is computed. The inverse one-dimensional transform would give back (b) from (c) or (e) from (f). But the final move is to take the Fourier transform of the two-dimensional set of data, built up from data like (c) and (f) (and (i) and (l) ...), to yield a two-dimensional representation of the original object. That this can be done is the gist of the **Fourier-slice theorem**.

Let's demonstrate the Fourier-slice theorem formally. We will assume that the original object is known, although in practical applications it is not known but rather what is to be determined. The projection of the object perpendicular to the x axis is

$$p(x) = \int_{-\infty}^{+\infty} \rho(x,y)\,dy. \tag{11.21}$$

Here ρ stands for the (optical) density of the material; it is summed up over the whole column representing all y values at a given x value to yield $p(x)$. We now compute the two-dimensional Fourier transform of ρ.

$$F(k_x, k_y) = \int_{-\infty}^{+\infty} \int_{-\infty}^{+\infty} \rho(x,y) \exp[-2\pi i(k_x x + k_y y)]\,dx\,dy. \tag{11.22}$$

(In practice, the data for F are built up by taking the one-dimensional Fourier transforms

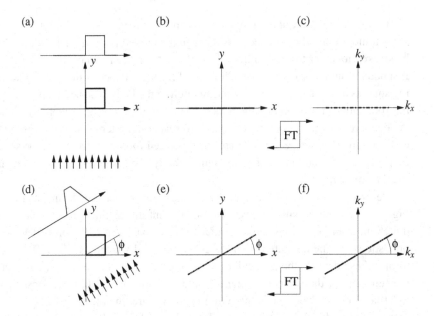

Figure 11.6 Projection tomography. The three columns represent real space, (a) and (d); projection space, (b) and (e); and Fourier space, (c) and (f). The terahertz radiation is represented by the lines with arrowheads. In (a), the object, represented by a square, is irradiated with a beam perpendicular to the x direction. The transmitted beam is detected on the other side; this is the *projection* perpendicular to x, $p(x)$, shown at the top of (a). In (b), the projection data $p(x)$ are shown lying along the x axis in the projection plane. In (c), the one-dimensional Fourier transform of $p(x)$ is shown in the Fourier plane. In (d), (e) and (f), the illumination has been rotated by an angle ϕ (here $\pi/6$) and the projection is now $p(x\cos\phi, y\sin\phi)$. The process is repeated for many angles of projection until a detailed image is built up in the Fourier plane. The *Fourier-slice theorem* states that the Fourier transform of the two-dimensional image thus built up in the Fourier plane reproduces the original object in real space.

of many one-dimensional projections.) Now let's consider a one-dimensional slice of the Fourier plane, specifically, along the k_x axis. Along the k_x axis, k_y is zero. So here

$$\begin{aligned}F(k_x, 0) &= \int_{-\infty}^{+\infty}\int_{-\infty}^{+\infty} \rho(x,y)\exp[-2\pi i(k_x x)]dxdy \\ &= \int_{-\infty}^{+\infty}\left[\int_{-\infty}^{+\infty}\rho(x,y)dy\right]\exp[-2\pi i(k_x x)]dx \\ &= \int_{-\infty}^{+\infty} p(x)\exp[-2\pi i(k_x x)]dx,\end{aligned} \qquad (11.23)$$

where, in the last step, Equation (11.21) was invoked. The final line is the one-dimensional Fourier transform of the projection along the x axis, as required.

A related mathematical entity is the **Radon transform**, which may be written as

$$R\rho(\phi, r) = \int_{-\infty}^{+\infty} \rho(x(s), y(s))ds. \qquad (11.24)$$

Here we are considering a particular ray which, at its closest, is a distance r from the origin; the perpendicular to the ray makes an angle ϕ with the x axis. The parameter

s runs along the ray from its origin (taken to be $-\infty$ here, although in practice it will be more, starting at the emitter) to its destination (taken to be $+\infty$ here, although in practice it will be less, ending at the detector). The Radon transform may be thought of as a plane-polar version of the Cartesian Fourier transform, more convenient to use in this experimental arrangement when the sample is rotated relative to the beam. (Figure 11.6 shows the sample stationary and the emitter and detector rotating; usually, the emitter and detector remain in place and the sample is rotated.)

The whole process may be moved up a notch. Rather than collect the original data as a set of one-dimensional projections, they may be collected as a set of two-dimensional projections. From these are computed two-dimensional Fourier transforms. These combined produce an image in three-dimensional space. From the inverse three-dimensional Fourier transform, the original three-dimensional object is retrieved.

So far in our discussion of tomography we have been employing geometric optics. What if diffraction comes into play? The problem of retrieving the sample geometry from the projection data becomes more difficult.

An overview of **diffraction tomography** is given in Figure 11.7. The panels have the same arrangement as in Figure 11.6. The effect of diffraction is to introduce curva-

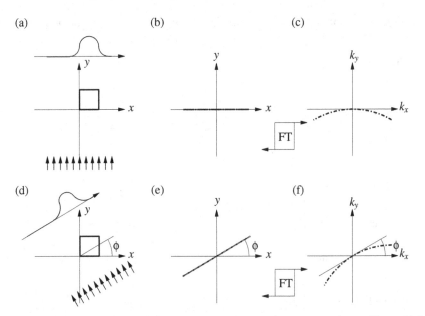

Figure 11.7 Diffraction tomography. The panels have the same meaning as Figure 11.6. Now diffraction effects are taken into account. In the real-space data, (a) and (d), sharp edges lose their definition. The experimental data are still plotted along straight lines in projection space, (b) and (e). These map to circles in Fourier space, (c) and (f).

ture to the formerly sharp shadows, as indicated in Figure 11.7a and Figure 11.7d. The projection now becomes

$$p(x) = p_0(x) + p_d(x), \tag{11.25}$$

where $p_0(x)$ is the diffraction-free component, as previously given in Equation (11.21), and $p_d(x)$ is the contribution due to diffraction. If we assume that the diffraction contribution is relatively small,

$$p_d(x) \ll p_0(x), \tag{11.26}$$

the **Rytov approximation**, the upshot is the slices in the projection plane map to arcs of circles in the Fourier plane. (I state this result without proving it.) As in the diffraction-free case, the inverse transform reproduces the original object.

11.6 Summary

11.6.1 Key terms

passive imaging, 234
angular resolution, 236
lateral resolution, 236
raster, 239
time-of-flight imaging, 241
focal plane array, 242
near-field imaging, 244
computer tomography, 247
diffraction tomography, 249

11.6.2 Key equations

f-number	$N \equiv \dfrac{L_f}{D}$	(11.1)
full cone angle	$\theta \sim \dfrac{D}{L_f} = \dfrac{1}{N}$	(11.2)
Rayleigh angle	$\theta_{\text{Rayleigh}} \sim \dfrac{1.22c}{fD}$	(11.6)
lateral resolution	$D_{\text{Rayleigh}} \sim \dfrac{1.22c}{f} N$	(11.8)
depth of focus	$\Delta_{\text{focus}} \sim \dfrac{2.44c}{f} N^2$	(11.10)
depth of field	$\Delta_{\text{field}} \sim \dfrac{2.44c}{f} \left(\dfrac{L_o}{D}\right)^2$	(11.11)
sub-wavelength aperture attenuation length	$\delta = \dfrac{a}{\sqrt{8\pi}}$	(11.20)

11.7 Table of symbols, Chapter 11

General mathematical symbols appear in Appendix B. If the unit of a quantity depends on the context, this is denoted '—'.

Symbol	Meaning	Unit
a	aperture of near-field mask	m
c	lightspeed	m Hz
d	depth of recess	m
D	aperture stop	m
D_{Rayleigh}	Rayleigh distance	m
E	electric field amplitude	V m^{-1}
E_0	electric field amplitude, initial	V m^{-1}
f	frequency	Hz
k	angular wavenumber	rad m^{-1}
k_x	x component of angular wavenumber	rad m^{-1}
k_y	y component of angular wavenumber	rad m^{-1}
k_z	z component of angular wavenumber	rad m^{-1}
L_f	focal length	m
L_o	object distance	m
n	refractive index	[unitless]
n_2	refractive index of dielectric	[unitless]
N	f-number	[unitless]
NA	numerical aperture	[unitless]
$p(x)$	projection (perpendicular to x)	—
p_0	projection without diffraction	—
p_d	additional projection term due to diffraction	—
r	distance from origin	m
s	length parameter	m
t	height of near-field mask	m
w	width	m
x	coordinate label	—
y	coordinate label	—
z	coordinate label	—
Δ_{field}	depth of field	m
Δ_{focus}	depth of focus	m
θ	full cone angle	rad
θ_{Rayleigh}	Rayleigh angle	rad
λ	wavelength	m
ρ	optical density	—
ϕ	angle of projection	rad

Appendix A Prefixes

Multiples			Submultiples		
Factor	Prefix	Abbreviation	Factor	Prefix	Abbreviation
10^{24}	yotta	Y	10^{-24}	yocto	y
10^{21}	zetta	Z	10^{-21}	zepto	z
10^{18}	exa	E	10^{-18}	atto	a
10^{15}	peta	P	10^{-15}	femto	f
10^{12}	tera	T	10^{-12}	pico	p
10^{9}	giga	G	10^{-9}	nano	n
10^{6}	mega	M	10^{-6}	micro	µ
10^{3}	kilo	k	10^{-3}	milli	m
10^{2}	hecto	h	10^{-2}	centi	c
10^{1}	deca	da	10^{-1}	deci	d

These are the prefixes defined in SI. Here are some rules for using them.

Writing prefixes
- use roman upright font
- use lower-case font, except make the first letter upper-case font when starting a sentence

Writing abbreviations of prefixes
- use roman font, except for µ
- use upright font – take care with µ
- use upper-case font for factors of 10^6 and greater
- use lower-case font for factors smaller than 10^6

Appendix B Mathematical symbols

B.1 Mathematical constants

Symbol	Meaning
e	the base of natural logarithms
i	the square root of negative one
π	ratio of the circumference to the diameter of the circle
∞	infinity

B.2 Mathematical unitary operations

Symbol	Meaning
$\sqrt{a}$	the square root of a

B.3 Mathematical relations

Symbol	Meaning
$a \equiv b$	a is defined to be equal to b
$a = b$	a is equal to b
$a \sim b$	a is approximately equal to b
$a \geq b$	a is equal to or more than b
$a > b$	a is more than b
$a \gg b$	a is much more than b
$a \leq b$	a is equal to or less than b
$a < b$	a is less than b
$a \ll b$	a is much less than b

B.4 Mathematical binary operations

Symbol	Meaning
$a + b$	b is added to a
$a - b$	b is taken from a
$a \times b$	a is multiplied by b
ab	a is multiplied by b
a/b	a is divided by b
$\dfrac{a}{b}$	a is divided by b
ab^{-1}	a is divided by b

B.5 Mathematical functions: trigonometry

Symbol	Meaning
e^x	e raised to the power of x
$\exp(x)$	e raised to the power of x
$\cos(x)$	cosine x
$\sin(x)$	sine x
$\tan(x)$	tangent x
$\arccos(x)$	angle whose cosine is x
$\arcsin(x)$	angle whose sine is x
$\arctan(x)$	angle whose tangent is x

B.6 Mathematical notation: complex numbers

Symbol	Meaning
$\tilde{z}$	$\tilde{z}$ is a complex number
$\tilde{z}^*$	the complex conjugate of $\tilde{z}$
$\mathrm{Re}(\tilde{z})$	the real part of $\tilde{z}$
z_1	the real part of $\tilde{z}$
z'	the real part of $\tilde{z}$
$\mathrm{Im}(\tilde{z})$	the imaginary part of $\tilde{z}$
z_2	the imaginary part of $\tilde{z}$
z''	the imaginary part of $\tilde{z}$
$\mathrm{mod}\ \tilde{z}$	the modulus of $\tilde{z}$
$|\tilde{z}|$	the modulus of $\tilde{z}$
$\arg(\tilde{z})$	the argument of $\tilde{z}$

B.7 Mathematical notation: vectors

Symbol	Meaning
a	**a** is a vector
â	**â** is a vector of unit length
a · **b**	the scalar product of **a** and **b**
a × **b**	the vector product of **a** and **b**

B.8 Mathematical operations: calculus

Symbol	Meaning
δx	an infinitesimal change in x
Δx	a small change in x
$\dfrac{dy}{dx}$	the differential of y with respect to x
$\dfrac{\partial y}{\partial x}$	the partial differential of y with respect to x
∇	the gradient operator, $(\partial/\partial x, \partial/\partial y, \partial/\partial z)$
$\int_a^b f(x)dx$	the integral of the function $f(x)$ with respect to x from a to b

Appendix C Mathematics

C.1 Circle

radius	r
circumference	C
diameter	D
pi	π
area	A

$$D \equiv 2r. \tag{C.1}$$

$$\pi \equiv \frac{C}{2r}. \tag{C.2}$$

$$A = \pi r^2. \tag{C.3}$$

C.2 Angle

radius	r
arc length	s
angle	θ

$$\theta \equiv \frac{s}{r}. \tag{C.4}$$

sense	
clockwise	−
anticlockwise	+

	Full circle	Half circle or straight angle	Quarter circle or right angle	Double circle
radians	2π	π	$\dfrac{\pi}{2}$	4π
degrees	360	180	90	720

C.3 Sphere

radius	r
surface area	A
volume	V

$$A = 4\pi r^2. \qquad (C.5)$$

$$V = \frac{4}{3}\pi r^3. \qquad (C.6)$$

C.4 Solid angle

radius	r
surface area	S
solid angle	Ω

$$\Omega \equiv \frac{S}{r^2}. \qquad (C.7)$$

	full sphere	hemisphere
steradians	4π	2π

C.5 Right-angled triangle

length of one side making right angle	x
length of other side making right angle	y
length of hypotenuse	r
angle opposite y	θ

C.5 Right-angled triangle

$$\sin\theta \equiv \frac{y}{r}. \tag{C.8}$$

$$\cos\theta \equiv \frac{x}{r}. \tag{C.9}$$

$$\tan\theta \equiv \frac{y}{x}. \tag{C.10}$$

$$[\text{Pythagoras}] \quad r^2 = x^2 + y^2 \tag{C.11}$$

$$r = \sqrt{x^2 + y^2}. \tag{C.12}$$

$$x = r\cos\theta. \tag{C.13}$$

$$y = r\sin\theta. \tag{C.14}$$

$$\theta = \arctan\frac{y}{x}. \tag{C.15}$$

$$\theta \equiv \arcsin\frac{y}{r}. \tag{C.16}$$

$$\theta \equiv \arccos\frac{x}{r}. \tag{C.17}$$

$\alpha =$	0	$\frac{\pi}{2}$	π	$\frac{3\pi}{2}$	2π	radians
	0	90	180	270	360	degrees
$\cos\alpha =$	1	0	−1	0	1	
$\sin\alpha =$	0	1	0	−1	0	

Mathematics

$$\cos^2 \alpha + \sin^2 \alpha = 1. \tag{C.18}$$

$$\cos(\alpha + \beta) = \cos \alpha \cos \beta - \sin \alpha \sin \beta. \tag{C.19}$$

$$\cos(\alpha - \beta) = \cos \alpha \cos \beta + \sin \alpha \sin \beta. \tag{C.20}$$

$$\cos \alpha + \cos \beta = 2 \cos \frac{\alpha - \beta}{2} \cos \frac{\alpha + \beta}{2}. \tag{C.21}$$

$$\cos(2\alpha) = \cos^2 \alpha - \sin^2 \alpha = 2 \cos^2 \alpha - 1. \tag{C.22}$$

$$\cos(-\alpha) = \cos(\alpha). \tag{C.23}$$

$$\sin(-\alpha) = -\sin(\alpha). \tag{C.24}$$

$$\cos(\alpha + \pi/2) = -\sin(\alpha). \tag{C.25}$$

$$\cos(\alpha + \pi) = -\cos(\alpha). \tag{C.26}$$

$$\sin(\alpha + \pi/2) = \cos(\alpha). \tag{C.27}$$

$$\sin(\alpha + \pi) = -\sin(\alpha). \tag{C.28}$$

$$\frac{d}{d\alpha} \cos \alpha = -\sin(\alpha). \tag{C.29}$$

$$\frac{d}{d\alpha} \sin \alpha = \cos(\alpha). \tag{C.30}$$

$$\int \cos \alpha \, d\alpha = \sin \alpha + C. \tag{C.31}$$

$$\int \sin \alpha \, d\alpha = -\cos \alpha + C. \tag{C.32}$$

$$\int \cos^2 \alpha \, d\alpha = \frac{1}{2}[\alpha + \sin(\alpha) \cos(\alpha)] + C. \tag{C.33}$$

$$\int \sin^2 \alpha \, d\alpha = \frac{1}{2}[\alpha - \sin(\alpha) \cos(\alpha)] + C. \tag{C.34}$$

C.6 Indices

$$a^m \times a^n = a^{m+n}. \tag{C.35}$$

C.7 Exponentials

$$\frac{d}{dx}e^x = e^x. \tag{C.36}$$

$$\int e^x \, dx = e^x + C. \tag{C.37}$$

$$e^{i\theta} = \cos\theta + i\sin\theta. \tag{C.38}$$

$$(\cos\theta + i\sin\theta)^n = \cos n\theta + i\sin n\theta. \tag{C.39}$$

Appendix D Further reading

The field of terahertz physics does not boast many elementary books. (I hope you have enjoyed this one.) Why not pursue the subject further through some of the recent books that take the subject to a higher level?

Sensing with Terahertz Radiation
Daniel Mittleman (Editor)
(Springer, 2003)
An edited collection of articles focussing on the detection of terahertz radiation.

Terahertz Optoelectronics
Kiyomi Sakai (Editor)
(Springer, 2005)
A book with much detailed technical information, featuring Japanese contributions to the field.

Terahertz Spectroscopy: Principles and Applications
Susan L. Dexheimer (Editor)
(CRC Press, 2008)
A book with much technical information, especially concerning spectroscopy.

Intense Terahertz Excitation of Semiconductors
S. D. Ganichev and W. Prettl
(Oxford University Press, 2006)
The subject matter is limited in both method (high intensity radiation) and the materials under investigation (semiconductors), but the treatment is definitive, even magisterial.

Principles of Terahertz Science and Technology
Yun-Shik Lee
(Springer, 2009)
A comprehensive overview citing much recent research literature.

Introduction to THz Wave Photonics
X.-C. Zhang and Jingzhou Xu
(Springer, 2010)
A volume summing up the recent research and with a strong emphasis on imaging.

Terahertz Techniques
Erik Brundermann, Heinz-Wilhelm Hübers and Maurice FitzGerald Kimmitt
(Springer, 2012)
A modernisation and amplification of Kimmitt's 1970 classic, 'Far-Infrared Techniques'.

Glossary

absolute value The magnitude of a quantity, separate to its direction or sign. 12

amplitude The size of an oscillation. The units depend on the the quantity being measured. For example, for electric field the units are V/m. 4

angular dispersion The change in angle relative to the change in frequency (or wavelength) in a grating spectrometer. 214

angular frequency The rate at which angle changes. 8

angular resolution The smallest angle an object can move perpendicular to the axis of an optical instrument for the change to be detected. 236

aplanatic lens A spherical lens sliced in such a way that all incoming rays focus to the same point. 176

argument The quantity that a function works on. 12

attenuated total reflectance Reflection at an angle that would usually result in total internal reflection but does not. 182

autocollimation The alignment of a blaze grating such that the returning beam lies along the same path as the incident beam. 214

band-pass filter A filter that passes frequencies within a certain range and stops frequencies below and above that range. 184

beam waist The narrowest part of a beam of light. 165

beam width The width of a beam of light, usually defined in terms of the distance from the axis of the beam corresponding to a given decrease in the field strength. 163

beat-frequency generation Generation of terahertz-frequency radiation by mixing together two higher-frequency signals. 149

beats Beats occur when two oscillations of the same amplitude but different frequencies are added together. In calling this phenomenon beats, it is assumed that the frequencies of the two original oscillations are similar. 31

blackbody An idealised object that completely absorbs all radiation at all frequencies. It is also a perfect emitter of radiation at all frequencies. 67

blackbody radiation Electromagnetic radiation from a blackbody. 67

blaze angle The angle of the ramps on a blazed grating. 213

blazed grating A diffraction grating comprising an array of shallow ramps. 213

bolometer A bolometer measures radiation by the change in electrical resistance of a cooled semiconductor. 198

chromatic resolving power The ability of a spectrometer to separate different colours (or frequencies). 215

circular polarisation A way in which light may be polarised. When viewed along the direction of the light propagation, the electric field vector moves in a circle. We may further distinguish right circular polarised light and left circular polarised light. 65

collimating lens A lens that takes diverging light from a point source and directs it into a parallel beam. 176

complex conjugate The partner of a complex number obtained by reversing the sign of the imaginary part. 13

complex number A number that may have a real part and an imaginary part. 12

complex plane A two-dimensional space in which a complex number can be represented. 12

computer tomography Tomography performed using a computer. 247

computer-aided tomography Tomography performed using a computer. 247

computer-axial tomography Tomography performed using a computer. 247

confocal The area near the beam waist, where the wavefront is close to planar. 165

continuously tuneable Able to be tuned to any desired value. 210

depth of field The distance an object can be moved without the focus of an optical instrument being changed appreciably. 238

depth of focus The distance a sensor can be moved at the focus of an optical instrument without the focus being changed appreciably. 238

detectivity The detectivity of a detector refers to the smallest signal that may be distinguished above the noise background. 192

diffraction grating A periodic structure designed or used for the purpose of diffracting light. 211

diffraction tomography Tomography in which diffraction effects are taken into account. 249

dipole radiation Electromagnetic radiation arising from the motion of two equal but opposite electrical charges. 140

discontinuously tuneable Able to be tuned, but only to certain discrete values. 210

dispersive spectroscopy Spectroscopy using a dispersive optical component such as a prism or a grating. 210

echelon When applied to a diffraction grating, refers to shallow ramps in a series. 213

electro-optic detector A detector that has at its heart an electro-optic crystal. Terahertz radiation falling on the crystal changes its properties in a way that is detected by a second, probe beam. The probe beam is usually near-infrared radiation. The plane of polarisation of the probe beam is rotated by the presence of the terahertz radiation. 200

electromagnetic wave A wave comprising inseparable electric and magnetic fields. The two fields are perpendicular to each other and perpendicular to the direction of propagation. The word 'light' is synonymous with electromagnetic wave in physics. 56

elliptical mirror A mirror in the shape of an ellipse. 173

elliptical polarisation A way in which light may be polarised. When viewed along the direction of the light propagation, the electric field vector moves in an ellipse. Linear polarised light and circular polarised light may be viewed as special cases of elliptical polarisation. 65

emission spectroscopy Spectroscopy of the light from a radiating object under study. 211

Euler's theorem A theorem relating the exponential function with the harmonic functions of cosine and sine. 14

evanescent An evanescent wave is one that does not propagate through a material but is strongly attenuated close to the surface. 181

f-number The ratio of the focal length to the entrance aperture of an optical instrument. 235

far field The area far from the beam waist, where the wavefront becomes spherical. 166

filters Devices that pass some frequencies and block others. 183

focal plane array An array of detectors near the focal point and perpendicular to the optical axis. 242

Fourier analysis To deconstruct a function into sine and cosine terms. 19

Fourier series The (possibly infinite) set of sine and cosine functions that appear in the Fourier theorem. 20

Fourier synthesis To construct a function by adding together sine and cosine terms. 19

Fourier-slice theorem Method of building a three-dimensional image by taking successive planar images. 247

Fourier theorem The statement that arbitrary functions can be constructed from a suitable set of sine and cosine functions. 19

free electron laser A laser that operates at terahertz frequencies based on the motion of electrons in a vacuum. 144

free particle A particle subject to no net potential. 77

free-standing grid A wire grid that is not supported by a substrate. 186

frequency The rate at which something takes place. *Units:* hertz. xiii, 4

frequency domain Presenting data with frequency as the variable. Showing values with frequency plotted on the horizontal axis. Compare and contrast with *time domain.* 6

frustrated total internal reflectance Reflection at an angle that would usually result in total internal reflection but does not. 183

fundamental frequency The lowest frequency supported by an oscillation of a given period. If the period is T, the fundamental frequency is $f = 1/T$. 19

geometrical optics Optics in which light travels in straight lines (or rays) until striking an optical element. 162

germanium laser A laser that operates at terahertz frequencies fabricated from the elemental semiconductor germanium. 146

Golay cell A Golay cell detects radiation through the expansion of a gas. 197
grating A periodic structure. 211

half-wave plate An optical device that retards one polarisation of light by half a cycle relative to the other polarisation of light. 187
harmonic Pertaining to circular motion. 9
hertz The unit of frequency. xiv
high-pass filter A filter that passes high frequencies and stops other frequencies. 184
higher order A diffraction grating order higher than the zeroth order. 213

imaginary number A number that, when squared, is negative. 12
initial phase The point on a cycle, measured in angle, at the starting time of a measurement. *Units*: radians. 5
interference The combining of two waves such that they add up (constructive interference) or cancel (destructive interference). 30
interferometry The method of spectroscopy that uses the interference of two beams of light as the working principle. 216
inverse metre The unit of spatial frequency. xv

Jones matrix The Jones matrix is used together with Jones vectors to describe the passage of light through an optical system. The output vector is obtained by multiplying the input vector by the Jones matrix. 50
Jones vector A simple way to represent the polarisation of light. 48

laser A device for producing coherent, monochromatic light. 140
lateral resolution The smallest distance an object can move perpendicular to the axis of an optical instrument for the change to be detected. 236
left circular polarisation A way in which light may be polarised. When viewed along the line of light propagation towards the light source, the electric field vector moves in a circle in an anticlockwise direction. Note, some authors use the opposite sense (clockwise). 65
lens An optical device changing the direction of rays of light passing through it. 174
linear polarisation A way in which light may be polarised. When viewed along the direction of the light propagation, the electric field vector moves in a straight line. Also known as plane polarisation. 65
linearity The linearity of a detector refers to how directly the output depends on the input. 191
low-pass filter A filter that passes low frequencies and stops other frequencies. 184

metre The unit of length. xv
mirror An optical component that reflects light. 166
modulus The magnitude of a complex number. 12
molecular laser A laser in which the lasing material is a molecular gas. 148

near-field The area near the beam waist, where the wavefront is close to planar. 165
near-field imaging Imaging based on detecting the light very close to the object. 244

nonlinear If two quantities, x and y, can be related by an equation of the form $y = mx + b$, they are said to have a linear dependence. If this equation does not describe their relation, they are said to have a nonlinear relationship. 158

optical path difference The equivalent distance light would travel in a vacuum to acquire the same phase change as it does through travelling a certain length in a material. 218

optical rectification A nonlinear optical process analogous to electrical rectification, producing the frequency difference of two relatively high frequency signals. 158

orthogonal At right angles to each other. The product of two orthogonal quantities is zero. 21

parabolic mirror A mirror in the shape of a parabola. 170

passive imaging Using light emitted by an object to image it. For example, taking a photograph of a star. Contrast with active imaging, where light is supplied as part of the imaging system. For example, flash photography. 234

periodic Occurring over and over again. Formally, a function f of time t, $f(t)$, is periodic with period T if $f(t) = f(t + T)$, for any value of t. Since the definition involves 'any value of t', the function must exist for all times, that is, for any t, $-\infty < t < +\infty$. Thus, by definition, a periodic function extends over all possible times; we cannot properly speak of a periodic function that began at 9 am and finished at 10 am. From the definition, we see the function takes on the same value at time $t + T$ as it has at time t; it has the same value at times $t + 2T$, $t + 3T$, $t - T$, among others. 4

phase The relative amount an oscillation is through its cycle. *Units:* radians. 9

photo-Dember emitter An emitter based on the photo-Dember effect. The 'photo' refers to the production of charge carriers by light. The 'Dember effect' refers to the different rates of diffusion of charge carriers of different signs, which leads to the production of a changing dipole. 157

photoconductive detector A photoconductive detector detects radiation by a change in the electrical conductivity of the device. 201

photoconductive emitter An emitter of terahertz-frequency radiation due to charge carriers, produced by pump radiation, being swept along by an applied electric field. 152

photon The smallest (minimum) amount of light. A photon has energy $E = hf = \hbar\omega$. A photon has momentum in the direction of its motion and of magnitude $p = \hbar k = h/\lambda = hf/c$. 55

plane mirror A flat mirror. 169

plane polarisation A way in which light may be polarised. The electric field vector lies in a plane. Also known as linear polarisation. 65

plane-polar coordinates Coordinates used to describe position in a plane relative to a centre or pole. One coordinate is the distance from the pole, the other is the angle made with a reference direction. 8

polarisation The systematic restriction of the electric field vector of a light wave to certain geometries. 38

polarisers Devices that cause an unpolarised beam of light to acquire polarisation. 186

polychromatic source A radiation source that emits several discrete frequencies of light simultaneously. 209

principle of superposition The principle that the net result of two waves in a medium is simply the addition of the effect of the two waves separately. 26

purely imaginary A complex number that has only an imaginary component and no real component. 12

purely real A complex number that has only a real component and no imaginary component. 12

pyroelectric detector A pyroelectric detector develops an electrical potential across its faces when radiation falls on it. 198

quantum cascade laser A laser consisting of a series of quantum wells. Electrons 'cascade' from one well to another, emitting terahertz-frequency radiation as they do. 147

quantum dot A device in which the allowed states are confined in all three spatial dimensions. 83

quantum well A device in which the allowed states are confined in one spatial dimension. 83

quantum wire A device in which the allowed states are confined in two spatial dimensions. 83

quarter-wave plate An optical device that retards one polarisation of light by one-quarter of a cycle relative to the other polarisation of light. 187

quasi-optics Optics that takes diffraction effects into account when drawing ray diagrams. 163

Radon transform Method of retrieving tomographic information. 248

rapid-scan interferometer An interferometer where many scans over the whole mirror scan range are added to produce the final interferogram. 219

raster To move from left to right, then slightly up or down, then repeat. 239

Rayleigh angle The angle subtended at an optical instrument corresponding to the Rayleigh criterion for resolution. 236

Rayleigh range The distance (in the direction of light propagation) from the beam waist in which the wavefront is close to planar. 165

real numbers A number that, when squared, is positive. 12

rectangular coordinates Coordinates used to describe position in a plane relative to two perpendicular axes. 8

right circular polarisation A way in which light may be polarised. When viewed along the line of light propagation towards the light source, the electric field vector moves in a circle in a clockwise direction. Note, some authors use the opposite sense (anticlockwise). 65

Rytov approximation Method of retrieving tomographic information in which diffraction effects are taken into account but considered to be small. 250

second Unit of time. xiv

sensitivity The sensitivity of a detector is the ratio of the output to the input. 191

spatial frequency The rate at which something takes place in space. *Units:* inverse metre. The unit of inverse centimetres, also known as wavenumber, is also in common use. 61

square well A potential well with abrupt changes at both ends. 80

step-scan interferometer An interferometer where a single scan, one step at a time, is made over the whole mirror scan range to produce the final interferogram. 219

surface-field emitter An emitter of terahertz-frequency radiation as a result of an electrical field near the surface of the material. 155

synchrotron A device that accelerates electrons around a circle to harvest the resulting electromagnetic radiation. 144

terahertz A million million hertz. xvi

thermal detector A detector that depends on the radiation raising its temperature. 195

thermocouple Comprises two dissimilar metals joined together and across which the electrical potential is measured. 196

thermopile Comprises a series of thermocouples. 196

time The inverse of frequency. xiv

time-bandwidth theorem The statement that the product of the duration of a pulse and the range of frequencies encompassed by the pulse has a minimum. Reducing the time will increase the bandwidth, and vice-versa. 16

time domain Presenting data with time as the variable. Showing values with time plotted on the horizontal axis. Compare and contrast with *frequency domain*. 6

time-domain spectroscopy Spectroscopy in which the raw data are collected in the time domain, then transformed to the frequency domain. 223

time-of-flight imaging Imaging based on the different times light takes in coming from different points on the object. 241

time-resolved spectroscopy Spectroscopy in which the spectra at various (usually very short) times are acquired. 230

uniform circular motion Motion along a circular path at a steady rate. 6

vacuum electronics The motion of electrons in free space (rather than in matter, the normal realm of 'electronics'). 129

vector A quantity that has not only a magnitude but also a direction. Moreover, the quantity must add according to the parallelogram rule of addition. This means that the components add independently and implies vector addition is commutative. 42

wave An oscillation in both time and space. The time component is characterised by its frequency (or, alternatively, by its angular frequency or by its period). The space component is characterised by its spatial frequency (or, alternatively, by its angular spatial frequency or by its wavelength). A wave is also characterised by its amplitude and by its initial phase. 60

Glossary

wave plates Devices that change the phase of one polarisation of light relative to the phase of the perpendicular polarisation. Examples are the quarter-wave plate and the half-wave plate. 187

wavelength The separation in space between points on a wave having the same phase. *Units:* metre. xv, 61

wavenumber A unit used in quantifying spatial frequency. It corresponds to the number of waves in a unit distance. xv, 61

zero path difference The arrangement of mirrors in an interferometer so that the two interfering beams travel exactly the same optical path length before recombining. 217

zeroth order Referring to the undiffracted beam from a diffraction grating. 213

Index

Page numbers in bold denote references to figures.

absolute value, 12, **13**
absorption, 98
absorption coefficient, 117
amplitude, 3, 4, **5**, 25
analysis, 207
angle, 6
angular dispersion, 214
angular frequency, **7**, 8
angular resolution, 236
aplanatic lens, 176, **177, 178**
Applications, 207–52
argument, 12, **13**
attenuated total reflectance, 182
attenuated total reflection, **182**
attenuators, 183
autocollimation, 214

band-pass filter, 184
Basics, 3–124
beam waist, 165
beam width, 163
beat frequency, 32
beat-frequency generation, 149
beats, 25, 26, 31–33
blackbody, 67
blackbody radiation, 54, **68, 69, 70**, 65–71
blaze angle, 213
blazed grating, 213, **213**
bolometer, 190, 198, **199**
Boltzmann constant, 66
bound states, 127, 140–44

calorimetric sensors, 190
chromatic resolving power, 215
circular polarisation, 25, 54, 65, **66**
collimating lens, 177, **177**
COMBINING OSCILLATIONS, 25–53
    compact notation for two-dimensional oscillations, 42–4
    counter-rotating circular oscillations, compact notation, 44–7
    Jones notation, 47–52
    oscillations in two dimensions, 36–47
    two counter-rotating circular oscillations, 41–42
    two perpendicular oscillations: different amplitude, 39–41
    two perpendicular oscillations: same amplitude, 36
    two oscillations in the same direction, 26–35
        two arbitrary oscillations, 26–7
        two identical oscillations, 27–8
        two oscillations that differ only in amplitude, 28
        two oscillations that differ only in frequency – beats, 31–3
        two oscillations that differ only in initial phase – interference, 29–31
        two oscillations that differ only in phase, 28–9
        two oscillations with nothing in common, 35
        two oscillations with only amplitude in common – more beats, 35
        two oscillations with only frequency in common – interference, 33–4
        two oscillations with only initial phase in common – beat-like, 34–5
compact notation, 11–15
compact notation for two-dimensional oscillations, 42–4
complex conjugate, 13
complex multiplication, **44**
complex number, 12, **13**
complex plane, 12, **13**
complex refractive index, 114
Components, 127–204
computer-aided tomography, 247
computer-axial tomography, 247
conductivity, 102
conductor, 108
    electromagnetic waves in, 113–18
confocal, 165
constant
    Boltzmann, 66
    electric, 57
    lightspeed, 58
    magnetic, 57
    Planck, 55

constructive interference, 31
continuous laser source, 149–51
continuously tuneable, 210
corner mirror, **170**
cos, **7**
cosine, **7**, 8, **20**
counter-rotating circular oscillations, compact notation, 44–7

damping force, 104
depth of field, 238
depth of focus, 238
describing electromagnetic waves, 62–4
describing oscillations, 4–11
describing waves, 60–62
detectivity, 192
DETECTORS, 190–203
   electro-optic, 200–1
   parameters, 191–5
   photoconductive, 201–2
   thermal, 195–200
dielectric breakdown, 101
dielectric function, 98, 101, **107–11**
dielectrics, 101
diffraction grating, 211, **211**, 213
diffraction tomography, 249, **249**
digital, 233
dipole radiation, 140
discontinuously tuneable, 210
dispersion, **78**, 207
dispersive spectrometer, 207
dispersive spectroscopy, 210–16

echelon, 213
efficiency, 208
electric constant, 57
electro-optic detector, 190, 200, **201**
electromagnetic wave, 54, 56–60, **62**
electromagnetic waves at an interface, 118–22
electromagnetic waves in matter, 101–4
electromagnetic waves in a conductor, 113–18
electromagnetic waves in an insulator, 112–13
ellipse, **38**, **40**, **47**
elliptical mirror, 173, **173**
elliptical polarisation, 25, 54, 65, **66**
emission spectroscopy, 211
energy
   thermal, 66
environmental factors, affecting detector, 192
Euler's theorem, 14
evanescent, 181
evanescent wave, **182**

f-number, 235
far-field, 166
filters, **184**, 183–5
finite potential well, 84–5
finite square well, **85**

focal plane array, 233, 242
focal plane array imaging, 242, **243**
Fourier analysis, 3, **18**, 19
Fourier coefficients, **18**
Fourier methods, 3
Fourier series, **18**, 20
Fourier synthesis, 3, **18**, 19
Fourier-slice theorem, 247
Fourier theorem, 17–23
Fourier transform spectrometer, 207
free charges, 127, 129–40
free electron laser, 127, 144
free particle, **78**, 77–80
free-standing grid, 186
frequency, xiii, 3, 4, **5**, 7, 25
frequency domain, **5**, 6, **17**, **18**, 19
frustrated total internal reflection, **182**, 183
function
   Planck, 67, **68–70**
   Rayleigh-Jeans, 68
   Wien, 67
fundamental frequency, 19

Gaussian beam, **164**, **166**
Gaussian beam propagation, 162
Gaussian optics, 162
Ge-laser, 127
geometrical optics, 162
germanium laser, 146
Golay cell, 190, 197, **197**
grating, 211
grating equation, 212

half-wave plate, 187
harmonic, 9
harmonic oscillation, **7**, 9
harmonic oscillator model, 104–12
hemispherical lens, **176**
hertz, xiv
high-pass filter, 184
higher order, 213

image, 232
imaginary axis, **13**
imaginary number, 12
IMAGING, 232–51
   focal plane array, 242
   near-field, 242–6
   passive, 234–39
   raster, 239–42
   reflection raster, 241–42
   tomography, 246–250
   transmission raster, 239–41
imaging
   reflection raster, 241–42
infinite potential well, 80–84
infinite square well, **84**
initial phase, 3, 5, **5**, 9, 25

insulator
    electromagnetic waves in, 112–13
INTERACTION OF LIGHT AND MATTER, 98–23
    electromagnetic waves at an interface, 118–22
    electromagnetic waves in matter, 101–4
    electromagnetic waves in a conductor, 113–18
    electromagnetic waves in an insulator, 112–13
    harmonic oscillator model, 104–12
    static electric field, 99–101
interface
    electromagnetic waves at, 118–22
interference, 25, 26, 29–31, 33–34, 207
interferogram, 207
interferometer, 207
interferometry, 216–22
INTRODUCTION, xiii–xix
inverse Boltzmann constant, 66
inverse metre, xv

Jones matrix, 50
Jones notation, 47–52
Jones vector, 48

laser, 140, 146–9
lateral resolution, 236
left circular polarisation, 65, **66**
lenses, 174–80
LIGHT, 54–73
    blackbody radiation, 65–71
    describing electromagnetic waves, 62–4
    describing waves, 60–62
    electromagnetic waves, 56–60
    photons, 55–6
    polarisation, 64–5
light
    visible, 56
lightpipes, 180
linear polarisation, 25, 65, **66**
linearity, 191
linearly polarised, 54
low-pass filter, 183

magnetic constant, 57
mathematical symbols [Appendix B], 254–6
mathematics [Appendix C], 257–61
MATTER, 74–97
    finite potential well, 84–5
    free particle, 77–80
    infinite potential well, 80–84
    multiple potential well, 85–7
    oscillations of molecules, 87–95
    Schrödinger equation, 75–7
    wave mechanics, 74–5
Maxwell's equations
    free space, 57

mechanics
    wave, 74–5
metre, xv
Michelson interferometer, 207, **217, 218**
mirrors, 166–73
modulus, 12, **13**
molecular laser, 127, 147, 148
molecules
    oscillations, 87–95
monochromatic sources, 208–10
monochromatic spectroscopy, **209**
multiple potential well, 85–7

near-field, 165
near-field imaging, 233, **245**, 242–6
nonlinear, 158
numerical aperture, 236

optical path difference, 218
optical rectification, 158
OPTICS, 162–89
    attenuators, 183
    filters, 183–5
    lenses, 174–80
        focussing, 175
        matching, 175
    lightpipes, 180–81
    mirrors, 166–73
        elliptical, 172
        parabolic, 170
        plane, 169
    polarisers, 186–7
        linear, 186
        wave plates, 187
    prisms, 180–3
    quasi-optics, 163–6
    waveguides, 180–81
    windows, 185–6
optics, 162
orthogonal, 21
OSCILLATIONS, 3–24
    compact notation, 11–15
    describing oscillations, 4–11
    Fourier theorem, 17–23
    time-bandwidth theorem, 15–17
oscillations, 3
oscillations in two dimensions, 36–47
    two counter-rotating circular oscillations, 41–2
    compact notation for two-dimensional oscillations, 42–4
    counter-rotating circular oscillations, compact notation, 44–7
    Jones notation, 47–52
    two perpendicular oscillations: different amplitude, 39–41
    two perpendicular oscillations: same amplitude, 36–9
oscillations of molecules, 87–95

Index

parabolic mirror, 170, **170–2**
parameters, 191–5
particle
  free, 77–80
passive imaging, **235**, 233–9
period, **7**
periodic, 4
phase, 9
phase difference, **30**
photo-Dember emitter, 157
photoconductive detector, 190, 201, **202**
photoconductive emitter, 151, 152
photons, 54–6
Planck constant, 55
Planck function, 67, **68–70**
plane mirror, 169, **169**
plane-polar coordinates, 8
plane polarisation, 65
plasma, 110
polarisation, 25, 26, 38, 54, 64–5
  circular, 25, 54, **66**
    left, **66**
    right, **66**
  elliptical, 25, 54, **66**
  linear, 25, 54, **66**
polarisation, of detector, 192
polarisers, 186–7
polychromatic source, 209
potential well
  finite, 84–5
  infinite, 80–4
  multiple, 85–7
prefixes [Appendix A], 253
principle of superposition, 25, 26
prisms, 181–3, **181**
projection tomography, **248**
pulsed lasers source, 151–60
purely imaginary, 12
purely real, 12
pyroelectric detector, 190, 198, **198**

quantum cascade laser, 127, 147
quantum dot, 83
quantum well, 83, **84**, **85**
quantum wire, 83
quarter-wave plate, 187
quasi-optics, **164**, **166**, 163–6

radiation
  blackbody, 54, **68**, **69**, **70**, 65–71
radius, 8
Radon transform, 248
range, of detector, 192
rapid-scan interferometer, 219
raster, 234, 239–42
ray optics, 162
Rayleigh angle, 236
Rayleigh range, 165

Rayleigh-Jeans function, 68
reading, further [Appendix D], 262
real axis, **13**
real numbers, 12
rectangular coordinates, 8
reduced mass, **89**
reflection, 98
reflection imaging, 233, 241, **241**
reflection raster imaging, 241–2
reflectivity, **167**
refraction, 98
refractive index, 112
relative dielectric function, 101
resolution, 208
right circular polarisation,
  65, **66**
Rytov approximation, 250

scattering, 98
Schrödinger equation, 74–7
second, xiv
sensitivity, 191
sensors, 190
  thermal
    bolometric, 198
    Golay, 197
    pyroelectric, 197
    thermocouple, 196
    thermopile, 196
sin, **7**
sine, **7**, 8, **20**
skin depth, **167**
SOURCES, 127–61
  bound states, 140–44
  driven by continuous lasers, 149–51
  driven by pulsed lasers, 151–60
  free charges, 129–140
  laser, 146–9
  survey, 128–9
  vacuum electronics, 144–6
sources
  monochromatic, 209–210
spatial frequency, 61
spectral range, 208
spectral response, 192
spectrometer, 207
SPECTROSCOPY, 207–31
  dispersive spectroscopy, 210–16
  emission, 213
  examples, 225
  interferometry, 216–22
  monochromatic sources, 209–210
  time-domain spectroscopy, 222–30
    compared to interferometry, 223
    configurations, 223
    time-resolved spectroscopy, 230
spectrum, 207

Index

speed of light in a vacuum, 58
speed, of detector, 192
spherical lens, **178**
square well, 80, **84, 85**
static electric field, 99–101
Stefan-Boltzmann law, 71
step-scan interferometer, 219
superposition, **27**
surface emitters, 155
   photo-Dember effect, 157
   surface field effects, 155
survey of sources, 128–9
symbols
   mathematical [Appendix B], 254–6
synchrotron, 127, 144
synthesis, 207

Terahertz
   applications, 207–52
   basics, 3–124
   components, 127–204
terahertz, xvi
thermal detectors, **195**, 195–200
thermal energy, 66
thermal sensor, 190
thermal sources, 127
thermocouple, 196
thermocouple detector, **196**
thermopile, 196
thermopile detector, **196**
time, xiv
time domain, **5**, 6, **17, 18**, 19
time-bandwidth theorem, 3, 16, **17**, 15–17
time-domain spectroscopy, 207, **225, 226, 228, 229**, 222–30
time-of-flight imaging, 241
time-resolved spectroscopy, 230
tomography, 233, 246–50
total internal reflection, **175**
transducer, 190
transmission, 98
transmission imaging, 233, 239, **240**
transmission raster imaging, 239–41
two arbitrary oscillations, 26–7
two counter-rotating circular oscillations, 41–2

two identical oscillations, 27–8
two oscillations in the same direction, 26–35
   two oscillations that differ only in amplitude, 28
   two oscillations that differ only in frequency – beats, 31–3
   two oscillations that differ only in initial phase – interference, 29–31
   two oscillations that differ only in phase, 28–9
   two oscillations with nothing in common, 35
   two oscillations with only amplitude in common – more beats, 35
   two oscillations with only frequency in common – interference, 33–4
   two oscillations with only initial phase in common – beat-like, 34–5
two perpendicular oscillations: different amplitude, 39–41
two perpendicular oscillations: same amplitude, 36–9

uniform circular motion, 6, **7**
unit circle, **10**
unit diamond, **10**
unit square, **10**

vacuum electronics, 129, 144–6
vector, 42
visible light, 56

wave, 54, 60
   electromagnetic, 54
wave equation
   one-dimensional, 61
wave mechanics, 74–5
wave plates, 187
waveguides, 180
wavelength, xv, 61
wavenumber, xv, 61
waves
   describing, 60–62
   electromagnetic, 56–60
Wien function, 67
Wien's displacement law, 71
windows, 185

zero path difference, 217
zeroth order, 213

Frequencies expressed in THz

	THz
Once in a ...	
year	0.00000000000000000003169
month	0.0000000000000000003803
week	0.000000000000000001653
day	0.00000000000000001157
hour	0.0000000000000002778
minute	0.00000000000001667
second	0.000000000001
kHz	0.000000001
MHz	0.000001
GHz	0.001
Hyperfine splitting in Cs^{133}	0.009192631770 [by definition]
THz	**1.**
Red light	400.
Green light	600.
Blue light	800.
Once in a ...	
atomic unit of time	41340.
natural unit of time	776300000.

THz photons

	Wavelength (in vacuum)	Wavenumber (in vacuum) cm^{-1}	Energy meV	Temperature K	Frequency THz
One Kelvin	14.39 mm	0.6950	0.08617	1	0.02084
	12.40 mm	0.8066	0.1	1.160	0.02418
One centimetre	10 mm	1	0.1240	1.439	0.02998
One wavenumber	10 mm	1	0.1240	1.439	0.02998
Helium boils	3.426 mm	2.919	0.3619	4.2	0.08751
	2.998 mm	3.336	0.4136	4.799	0.1
	1.439 mm	6.950	0.8617	10	0.2084
One meV	1.240 mm	8.066	1	11.60	0.2418
One millimetre	1 mm	10	1.240	14.39	0.2998
	999.3 µm	10.01	1.241	14.40	0.3
One THz	**299.8 µm**	**33.36**	**4.136**	**47.99**	**1**
Nitrogen boils	185.9 µm	53.80	6.670	77.4	1.613
	143.9 µm	69.50	8.617	100	2.084
	124.0 µm	80.66	10	116.0	2.418
	100 µm	100	12.40	143.9	2.998
	99.93 µm	100.1	12.41	144.0	3
Triple point	52.67 µm	189.9	23.54	273.16	5.692
Room temperature	49.08 µm	203.7	25.26	293.15	6.108
	47.96 µm	208.5	25.85	300	6.251
	29.98 µm	333.6	41.36	479.9	10
	12.40 µm	806.6	100	1160	24.18
	10 µm	1000	124.0	1439	29.98
One micrometre	1 µm	10000	1240	14388	299.8

Exact relationships in SI units

$$f = c/\lambda \qquad \lambda = c/f$$
$$f = c\sigma \qquad \sigma = f/c$$
$$f = E/h \qquad E = hf$$
$$f = k_B T/h \qquad T = hf/k_B$$

Exact relationships in convenient units

f [THz] = 299.792458/λ [µm] λ [µm] = 299.792458/f [THz]
f [THz] = 0.0299792458σ [cm^{-1}] σ [cm^{-1}] = f [THz]/0.0299792458

Approximate relationships in convenient units

f [THz] = 0.2417989348 × E [meV] E[meV] = 4.135667516 × f [THz]
f [THz] = 0.020836618 × T [K] T[K] = 47.992434 × f [THz]

Visible phenomena expressed in THz

Phenomenon	nm	THz
Proposed low-frequency limit		400
Conventional long-wavelength limit	700	428
He-Ne laser	633	474
Na-D line	589	509
Conventional short-wavelength limit	400	749
Proposed high-frequency limit		800

Semiconductor bandgaps (300 K) expressed in THz

Semiconductor	eV	THz
InSb	0.17	41
InAs	0.35	85
Ge	0.66	160
GaSb	0.73	176
Si	1.12	271
InP	1.34	324
GaAs	1.42	343
CdTe	1.49	360
CdSe	1.74	421
GaP	2.26	546

Phonon frequencies expressed in THz

Compound	TO phonon THz	LO phonon THz
NaCl	4.9	8.0
InAs	6.5	7.2
GaAs	8.2	8.8
InP	9.1	10.3
GaP	11.0	12.1

Key equations

harmonic oscillation	$A = A_0 \cos(2\pi ft + \delta) = A_0 \cos\phi$
time-bandwidth theorem	$\Delta f \Delta t \geq 1$
Fourier theorem	$g(t) = a_0 + \sum_{n=1}^{\infty} [a_n \cos(2\pi n f_0 t) + b_n \sin(2\pi n f_0 t)]$
interference	$A_1 + A_2 = 2A_0 \cos\left(\frac{\delta_1 - \delta_2}{2}\right) \cos\left(2\pi ft + \frac{\delta_1 + \delta_2}{2}\right)$
beats	$A_1 + A_2 = 2A_0 \cos[2\pi \frac{f_1 - f_2}{2} t] \cos[2\pi \frac{f_1 + f_2}{2} t + \delta]$
Planck-Einstein relation	$E = hf = \hbar\omega$
general harmonic wave	$A_0 \cos(kx - 2\pi ft - \delta) = A_0 \cos\phi$
Planck function	$B_f(T) = \frac{2hf^3}{c^2} \left[\exp\left(\frac{hf}{k_B T}\right) - 1\right]^{-1}$
time-independent Schrödinger equation	$\frac{d^2 \psi(x)}{dx^2} = \frac{2m}{\hbar^2}[U(x) - E]\psi(x)$
free-particle wavefunction	$\psi(x) = \sqrt{\frac{2}{L}} \sin(kx)$
energy of particle in a well	$E_n = \frac{\hbar^2 k_n^2}{2m} = n^2 \frac{h^2}{8mL^2}$
energy of rotational states	$E(J) = \frac{\hbar^2}{2I} J(J+1)$
energy of quantum harmonic oscillator	$E(n) = (n + \frac{1}{2})\hbar\omega = (n + \frac{1}{2})hf$
plasma frequency	$\omega_p^2 \equiv \frac{nq^2}{\epsilon_0 m}$
effective dielectric function	$\tilde{\epsilon} = \epsilon_0 \left[\frac{\omega_p^2}{\omega_0^2 - \omega^2 - i\omega\gamma} + 1\right]$
effective conductivity	$\tilde{\sigma} = \epsilon_0 \omega \left[\frac{\omega_p^2}{\omega\gamma + i(\omega_0^2 - \omega^2)} - i\right]$
refractive index	$n(\omega) \equiv \frac{c}{v}$
relative dielectric function	$\epsilon_r(\omega) = n^2(\omega)$
absorption coefficient	$\alpha(\omega) = 2n_2(\omega)\frac{\omega}{c}$

Printed in the United States
By Bookmasters